Bristol Studies in International Theory

Series editors: **Felix Berenskötter**, SOAS, University of London, UK, **Neta C. Crawford**, Boston University, USA and **Stefano Guzzini**, Uppsala University, Sweden, PUC-Rio de Janeiro, Brazil

---

This series provides a platform for theoretically innovative scholarship that advances our understanding of the world and formulates new visions of, and solutions for, world politics.

Guided by an open mind about what innovation entails, and against the backdrop of various intellectual turns, interrogations of established paradigms, and a world facing complex political challenges, books in the series provoke and deepen theoretical conversations in the field of International Relations.

Also available

The Idea of Civilization and the Making of the Global Order
By **Andrew Linklater**

What in the World?
Understanding Global Social Change
Edited by **Mathias Albert** and **Tobias Werron**

Coming soon

Praxis as a Perspective on International Politics
Edited by **Gunther Hellmann** and **Jens Steffek**

Broken Solidarities
How Open Global Governance Divides and Rules
By **Felix Anderl**

Pluriversality and Care
Rethinking Global Ethics
By **Maggie FitzGerald**

Find out more

bristoluniversitypress.co.uk/
bristol-studies-in-international-theory

Bristol Studies in International Theory

Series editors: **Felix Berenskötter**, SOAS, University of London, UK, **Neta C. Crawford**, Boston University, USA and **Stefano Guzzini**, Uppsala University, Sweden, PUC-Rio de Janeiro, Brazil

---

Find out more

bristoluniversitypress.co.uk/
bristol-studies-in-international-theory

SNAPSHOTS FROM HOME

Mind, Action and Strategy in an Uncertain World

K.M. Fierke

BRISTOL
UNIVERSITY
PRESS

First published in Great Britain in 2022 by

Bristol University Press
University of Bristol
1–9 Old Park Hill
Bristol
BS2 8BB
UK
t: +44 (0)117 954 5940
e: bup-info@bristol.ac.uk

Details of international sales and distribution partners are available at bristoluniversitypress.co.uk

British Library Cataloguing in Publication Data
A catalogue record for this book is available from the British Library

ISBN 978-1-5292-2261-6 hardcover
ISBN 978-1-5292-2262-3 paperback
ISBN 978-1-5292-2263-0 ePub
ISBN 978-1-5292-2264-7 ePdf

Cover design: blu inc, Bristol
Front cover image: Unsplash/Yeshi Kangrang
Bristol University Press uses environmentally responsible print partners.
Printed in Great Britain by CMP, Poole

In loving memory of my beautiful Mia

Contents

List of Figures and Tables

Figures

Table

Acknowledgements

I would like to give special thanks to the Leverhulme Trust for their generosity in providing me with a fellowship to write this book, and, in light of the unusual circumstances in which it was written, for also agreeing to allow me to use my unused conference budget to employ three PhD research assistants to provide feedback on the manuscript. Ahmed Abozaid and Chaeyoung Yong, who read the first draft, and Shambhawi Tripathi, who read both the first and the almost final draft, provided invaluable comments which strengthened the text considerably. I would also like to thank Francisco Antonio-Alfonso and Dale Fierke, both of whom also provided insightful comments on the first draft, the editors and production staff of the Bristol International Theory series, the reviewers of the manuscript, Yang Yuanfuyi for her help with the Chinese characters in Snapshots 3 and 4, and Mary Jane Fox for doing the index. The students who over several years participated in my MLitt module at the University of St. Andrews, Agency and Strategy in Non-Western Political Thought, contributed to my thinking about the topic in the early stages, as did a number of people who read and commented on the fellowship application, individual chapters or articles that informed the argument, including Thierry Balzacq, Roland Bleiker, J. Peter Burgess, Michael Dillon, Vivienne Jabri, Tony Lang, Nicola Mackay, Philipps O'Brien, Laura Sjorberg, Michael Schnabble, Gabriella Slomp, Nadine Voelkner, Alexander Wendt and Laura Zanotti. Others, perhaps without knowing it, were a source of insight and support over the years or as the book was being written during lockdown, including David Fierke, Kathryn Follon, Sue Jenkins and Mary Stevens. I would also like to acknowledge Lily Ling, who, prior to her tragic death, was an inspiration through both her work and our conversations at conferences and during her visits to St. Andrews. Given their constant companionship during multiple lockdowns, I must further thank my collies: Trixie, who left me at age seventeen, just after the first lockdown ended, and Mia, who found her way to me as the second was beginning.

All books build on thought that preceded them, and in this respect citations are an important form of thanks. A larger number of scholars from international relations and beyond could have been cited in regards to the vast array of topics explored in this book, and some may wonder about their absence. The analysis that follows is an attempt to step out of more familiar concepts and reposition the apparatus to engage with the world from a different angle, and indeed, as one reviewer commented, this is *not* a conventional international relations text. In working across a very interdisciplinary field, I have had to limit my engagement with scholars closer to my academic home. I have nodded to related literatures, but the more in-depth explorations focus on concepts from Daoism, Buddhism and Hinduism and the parallel, drawn by a number of quantum physicists, to quantum physics. There are further parallels that could be drawn between the subject matter explored here and existing scholarship in international relations, Western social theory and thought systems from other global locations, which I hope will be a fruitful area for ongoing conversation that extends beyond the remit of this book.

Introductions

Repositioning the Apparatus

> Truth is one; people call it by different names.
>
> *Rig Veda*, Samhita 1.164.46

In 2020 uncertainty became the new normal as the world was gripped by pandemic, financial instability, extreme weather 'events' and political polarization, among others. The unpredictability of rapid changes with global impact highlighted the difficulty of decision-making and action in conditions of radical uncertainty. What happens to agency in conditions that are so unpredictable, where conventional wisdom can within a matter of weeks, or even days, be turned on its head in response to a dramatically changing context? This book approaches the problem through an exploration of a very old parallel between quantum physics and Daoism, Buddhism and Hinduism. Over the past one hundred years physicists have drawn on this parallel for a number of different reasons. In revisiting the parallel, I seek to explore what it might tell us about mind, action and strategy in a radically uncertain world. The question is timely in light of the dramatic changes brought about by global upheavals, from climate change to the COVID-19 pandemic.

Change has been a consistent feature of global life, even if this has not always been recognized by those who study it. As Nassim Taleb (2010) notes, people often act from an assumption of certainty, yet history is driven by uncertainty. The upheavals of 2020 were accompanied by an unusual degree of uncertainty. But the pandemic is just the most recent of a series of global changes. As Katzenstein and Seybert (2018: xi) state, the world has been persistently 'blindsided by the unexpected.' The end of the Cold War or the attacks on the Twin Towers on 11 September 2001, among others, ushered in significant global changes. Change is arguably the rule rather than the exception.

Earlier debates in international relations revolved around a question of whether change is possible at the international level. The contours of the current conversation look much different, although cut from this earlier cloth. This project is situated at the intersection of two debates that transcend disciplinary boundaries and would seem on the surface to have little to do with one another: the relevance of quantum theory for the social sciences and the importance of moving toward a global IR. The parallel between quantum theory and the Eastern philosophies brings the intersections between these debates to light.

Quantum social science

The first debate regards the implications of quantum theory for the social sciences and whether it can ever be anything more than pure theory or analogy. How one applies the theory is not entirely clear (Lamb-Books, 2016), not least because quantum effects are said to wash out once one moves above the subatomic level (Waldner, 2017). The question, as Murphy (2020: 11) notes, is whether the quantum project is a 'a mere fetishism of science, a positivist project lacking in laboratory evidence or an unnecessary import of complicated terminology to describe mechanisms and relations that can be adequately described in more established social scientific or theoretical language.'

The 'quantum turn' has the potential to shift the debate about science away from the hierarchical construction that pits positivists against 'others' and instead raises questions about the relationship between the two sciences and their respective usefulness for understanding the social and political world. Most interventions in this debate, which has steadily gained momentum over the last decade, have questioned the metaphysics of science that underlie claims about 'reality'. Karen Barad (2007) and Alexander Wendt (2015) each made a case regarding the relevance of quantum physics for the social sciences. Both authors focus on debates within the sciences, although Barad's work has had a significant impact on feminist and post-humanist theorizing. Both claim that quantum states are macroscopic, which refers to phenomena that can be observed with the naked eye, as well as microscopic, or that which can only be seen with a microscope. Neither goes far enough in elaborating the significance of these arguments for what social scientists do.

Further interventions have highlighted the resonance of quantum theory with critical theory.[1] Laura Zanotti (2019) draws on Barad and Foucault to explore the implications of ontological entanglement for how agency and

[1] For purposes of specification, my use of 'quantum turn' refers to the shift from a Newtonian to a quantum framework. Quantum theory refers to the theoretical basis of quantum physics. Quantum physics and quantum mechanics both refer to the study of sub-atomic particles, but the mechanics is more specifically about the formulation of mathematical law. As will become clear, my more frequent reference to the physics, as distinct from the theory or the mechanics, is a reflection of the language used by physicists who articulated the parallel between quantum physics and ancient Asian philosophies. Wendt (2015) focuses, by contrast, on quantum theory, and its underlying metaphysics.

ethics are conceptualized. She argues for the need to move toward a non-Newtonian ontology. The concern is echoed by Michael Murphy (2020), who looks to the parallel between quantum social theorizing and critical approaches, making the case for an alliance between them. But even critical theorists are not entirely convinced. For instance, Laura Sjorberg (2020) expressed ambivalence about the added value of a quantum turn given that feminists and others have long been making similar ontological claims. A further strand, represented by Bentley Allan (2018a) and Milja Kurki (2020), explores the implications of a relational cosmology for understanding international relations. At the core of the debate is a question about the added value of a quantum turn and what it would mean in practice, not only for the social sciences, and specifically international relations (IR), but for the relationship between the social and natural sciences. A special issue of *Security Dialogue*, edited by DerDerian and Wendt (2020), targeted the question of application, asking what it would mean to quantize IR, with separate articles exploring the significance of this for systems theory, technology, economics, memory and consciousness.

The social sciences have not been quick to embrace a quantum turn.[2] Yet, in a context characterized more by uncertainty than stability, there would seem to be a clear rationale for doing so. There is evidence of an increased interest in exploring how power works in conditions of uncertainty (Hymans 2020). Power, in this conceptualization, is less about control than 'the effect of actors' improvised and innovative responses to an incalculable environment or their experience of the world as equally uncertain' (Katzenstein and Seybert, 2018: 10–11). The Project Q symposia, organized by James DerDerian at the University of Sydney, has brought together a broad interdisciplinary group to explore the potentials of a quantum social science. Alexander Wendt organized the first 'quantum bootcamp' in July 2021. Momentum is developing but there remain many unanswered questions about the implications of a shift from assumptions of materiality, mechanism, determinism and localism, which Wendt (2015) argues have underpinned the social sciences, to those of consciousness,

How these distinctions are understood relate to large debates about the nature of reality, not least as it relates to language, which will be explored throughout the book.

2 As a debate across the social sciences, there are a variety of further intersections and conversations that span in numerous directions, including the application of principles from quantum mechanics to decision-making in psychology and the modelling of finance and economics (Haven and Krennikov, 2013); explorations into the intersections between quantum physics and indigenous thought (Peat, 2002; Bowman, 2019); environmental ethics (O'Brien, 2016, Hamilton, 2017); actor-network theory, which is informed by science and technology studies (see, for example, DeGoede and Sullivan, 2016; Salter 2015, 2019) and arguments that build on Deleuze's concept of an assemblage (Abrahamsen and Williams, 2011; Salter, 2013; Bueger, 2018).

non-locality, indeterminism and non-linearity for the study of human, social and political phenomena.

The added value of looking for answers in ancient thought may seem less obvious, yet the uncertainty at the heart of both past and present means that probability and potentiality are more the issue than predictability. Buddhism, Daoism and Hinduism provide a reservoir of knowledge, developed over millennia, related to questions of mind, action and strategy in conditions of uncertainty. Looking at specific points of parallel to quantum physics may provide insight into what it means for humans to live in a quantum world. Finally, as I will shortly discuss, the past is already a part of the present. Many practices associated with these traditions still have a place, not only in Asia but in the West as well.

Global international relations

On the one hand, the parallel between quantum physics and Daoism, Buddhism and Hinduism provides an opening to explore the macroscopic implications of a quantum turn. On the other hand, it may assist in addressing a disciplinary blind spot regarding the potential for a more global IR. The blind spot is a byproduct of the roots of IR in the United Kingdom and United States and its indebtedness to a particular understanding of science. Global IR seeks to reach out to the larger world, to voices and traditions that have been largely excluded from reflection on the nature of international or global politics. While a discipline that seeks to say something about practice within a space that covers the globe would seem to be global by definition, IR, as a field of study, emerged in the wake of World War I, at the heart of British empire, and later travelled to a new centre, both academic and political, in the United States following World War II. The discussion of what IR is, how it should be theorized and how it should be researched has been heavily localized in the US and UK, which have been centres of global power relations for the last few centuries.

Post-colonial debates and the call for a global IR have highlighted the ethnocentrism of the discipline and its concepts, given their origin in the West (for example, Bilgin, 2010, 2016; Shilliam, 2011; Jabri, 2012; Acharya, 2014; Epstein, 2017). One consequence of this ethnocentrism has been an absence of attention to other voices and systems of knowledge, many discredited or silenced with imperial expansion, in shaping how a purportedly global phenomenon is understood. A decade ago, Acharya and Buzan (2010) raised a question of why there is no non-Western international theory. Acharya's (2014) further articulation of a concept of 'Global IR' represents an attempt to approach history, practice and theory from a more global angle that moves beyond the West.

Questions about what thought systems from beyond the West might contribute to IR are far from new. Some twenty-five years ago Roland Bleiker (1993) drew on ancient Chinese philosophy to explore neorealist claims, and somewhat later, Chan, Bleiker and Mandaville (2001) pointed to the limited geographic and cultural space from which the discipline emerged, stating that most IR scholars wouldn't even know how to ask a question of agency, karma or fate from the perspective of Hindu or Buddhist cosmologies, or, one might add, that of other non-Western cosmologies. More recently, L.H.M. Ling (2013, 2014) developed the implications of Daoism for the conceptualization of global politics, while Mustapha Pasha (2017) has explored the relevance of Islam for international theory. Debates in China, among other places, have raised questions about the potential contribution of more ancient systems of thought, such as Confucianism, to a rethinking of international relations theory (Zhao, 2009; Callahan and Barantbatseva, 2012; Zang and Chang, 2016). Qin Yaqing (2016, 2018) contrasts a Confucian/Daoist concept of 'relationality' with the rationalism of IR theory (see also Kavalski, 2017; Nordin et al, 2019). Shi Chih-Yu et al (2019) proposed a balance of 'relationality' as an alternative to the realist balance of power. Relationality rests on assumptions that resonate with those of quantum entanglement (see, for example, Kurki, 2020; Pan, 2020).

The problem that has dogged this debate concerns the difficulty of getting beyond a particular conception of the relationship between science and culture, a topic which is explored in more depth in the next chapter. Science is understood to be universal, in contrast to the particularity of culture. Approaches that build on systems of thought beyond the West are often dismissed, either because they present ideas that can already be found in Western philosophy or social theory and which are thus said to be unnecessary (Wang, 2013: 35), or because they are associated with culture and thus have nothing to say about a more universal 'reality'. The debate over what international theory beyond the West might look like reveals an interesting tension between science and culture. Qin Yaqing (2007), prior to Buzan and Acharya's article on non-Western theory (2010), asked why there is no Chinese international theory. Some have argued that non-Western IR theory might arise from theories with distinct Russian, Chinese, Korean or Brazilian 'characteristics' (Song, 2001; Kim, 2016). Other Asian scholars of a more positivist bent have expressed concern that theory with characteristics is not sufficiently scientific (Qin, 2011; Eun, 2018) and expresses not only cultural but national qualities.[3] The debate is interesting

[3] With the publication of Yan Xuetong's (2011) edited collection in English by Princeton University Press, Chinese international theory arrived on the global map, but heavily influenced by a 'scientific method' involving quantitative techniques and indebted to positivism (Zhang, 2012). Building theory on 'characteristics' of a cultural origin highlights

and highlights the perplexing relationship between science and culture. It further raises a question of whether it is enough to replace one ethnocentric theory with another.

Western science is elevated above other traditions, which are assumed to be based on culture and lacking in systematicity. Within the field of IR, the West/non-West hierarchy overlaps with that between positivists or rationalists and 'interpretivists' or 'reflectivists',[4] from which science claims the high ground (for example, King et al, 1994) and any form of inquiry that does not conform to this unitary model is considered non-science. Such claims neglect the extent to which Western science is a byproduct of the interaction and exchange between Europe and other cultures (Hobson, 2004). They also neglect the interface between Western science and Western culture.

Cultural appropriation

Uncertainty has been the historical experience of populations who were subjected to the global project of Western expansion. The division of the world and the construction of sovereign boundaries did not correspond with more indigenous cultures, languages and practices, in either spatial or metaphysical terms, and indeed often resulted in the silencing or suppression of indigenous knowledge, as cultures were destroyed or dramatically transformed (Kilomba, 2010; Tuhiwai Smith, 2012; Ndlovu-Gatsheni, 2018). In so far as the 'civilizing mission' was justified by the superiority of Western 'science', which often provided evidence of racial superiority and inferiority, an intersection between the two debates begins to surface.

The parallel between quantum physics and Asian philosophy provides a particular cut into the problem that potentially addresses weaknesses at the heart of each debate. The cut also intersects with a further social science literature that raises a somewhat different concern about culture; that is, that the West has culturally appropriated a range of practices from the East (see, for example, Hsu, 2016; Mitra and Greenberg, 2016; Titmuss, 2016; Walsh, 2016), in which process their original meaning and significance has been 'lost in translation' (Compson, 2014). While the presence of the past has many different meanings, in this context, practices with origins in Buddhism, Daoism and Hinduism have travelled and taken root in the West, just as modern capitalism and science have reverberated out across the globe. Practices such as mindfulness, yoga, reiki or tai chi, to name just a few, have

---

the importance of positionality. However, positivist methods rest on assumptions that are contrary to older scientific traditions and philosophies with origins in Asia, which provide a different take on the universe or 'reality' that runs parallel to quantum physics.

4 The rationalist/reflectivist distinction was framed by Keohane (1988).

acquired an institutional place in Western culture, from schools to prisons, the military and even parliaments. The multi-directional movements have given rise to concerns about the dangers of cultural appropriation, while also highlighting a question of what happens when diametrically opposed assumptions meet in a particular context of practice.

Given the greater difficulty of visiting either gyms or doctors in a context of pandemic and sustained lockdown, the practices mentioned in the previous paragraph became even more important to some. Health and well-being are not usually associated with international politics, although, as Howell (2014) notes, medicine and war have developed side by side. While Howell's focus is the West, the same is true in the East. Asian philosophies of mind, action and strategy gave rise to practices relevant not only to the mind and body but to politics and war as well. The historical and symbiotic relationship between practices of health and war provides a context for exploring the implications of the parallel between quantum theory and the Eastern philosophies for understanding global politics.

The aforementioned parallel, also referred to as 'the parallel', makes it possible to rethink the lines of these debates (Fierke, 2017, 2019). The grounding of concepts in some notion of science is important, not because of science envy but rather to counter the prevailing certainty that because the world works in a particular way we must necessarily act in a particular way, which in IR has usually rested on assumptions of determinism, materialism and egoism. If 'reality' is indeterminate as well as relational and entangled, and not only at the microscopic level, then our assumptions about both theory and practice need to be rethought. Quantum impermanence, complementarity and entanglement – the three points of parallel – provide a backdrop for rethinking the nature of mind, action and strategy in an uncertain world. The Asian philosophies provide a seamless transition between quantum science and macroscopic patterns and forms of life. Both quantum physics and the Eastern philosophies go beyond the binary of analogy or actuality that has framed the social science debate (Murphy, 2020: 44) towards a conception life and world as both real and non-real. As such, contradiction is central to the dynamics of each.

The parallel points to philosophies and scientific practices that have for millennia relied on insights more recently rediscovered by quantum physicists. It provides an apparatus for beginning to unpack the implications of a quantum turn for macroscopic relations, from nature to social life to global politics. Both quantum physics and the ancient traditions shift away from the classical focus on the material properties of ontologically separate objects or individuals, that is, their 'thing-ness', toward an epistemology/ontology of entangled relations, within which polar opposites are mutually implicated, and from which difference emerges out of a dynamic process of mutual constitution. The Eastern philosophies not only lend further substance to Wendt's (2015: 260) claim that 'mutual constitution' is untenable from the perspective of a Newtonian metaphysics, Buddhism, Daoism and Hinduism also represent a particular angle from which to view how mutual constitution works in the world.

The apparatus

The parallel provides an apparatus for making a particular cut into questions of mind, action and strategy in an uncertain world. An apparatus is a measuring device. In classical physics, the apparatus and the scientist are separate from the objects of measurement. In a quantum world of uncertainty, the object of measurement is not fixed and what is seen depends heavily on how the apparatus is positioned. As Karen Barad (2007: 148) states, 'Apparatuses enact agential cuts that produce determinate boundaries and properties of "entities" within phenomenon.' As such the apparatus is a crucial part of an experiment. The choice of apparatus shapes the angle of observation. 'I' as analyst do not assess a reality that exists independently of me; rather what is seen is impacted by my position, which is also not fixed but entangled. To reposition the apparatus is thus to observe a phenomenon from a different angle. The apparatus in this case is the parallel between quantum impermanence, complementarity and entanglement and counterparts in ancient Eastern philosophy.

Daoism and Buddhism, no less than quantum theory, highlight the importance of the perspective of the observer. The concept of positionality, which has long been a concern of feminist and critical scholars, expresses a similar point regarding the importance of position and experience for how the world is perceived and how one is perceived within the world. Positionality can refer to how one's position or identity impacts on one's reception by others, what the researcher sees and how the story is told.[5] But it also assumes an incomplete view, which brings with it certain biases. Positionality comes with blind spots that may make it difficult to 'see' a complete view of reality.

The events of 2020 have been experienced globally, but from different positions in social, economic, cultural and geographical space. Whatever one's location, the blinders were peeled away in layers, making it possible to look at 'normal' life from a different angle. Through months of total or partial lockdown, against the backdrop of pandemic, we all occupied a liminal place between a past which in hindsight seemed a bountiful place of wealth and freedom, at least for some, and a future that offered only disease,

5 On positionality in a context of fieldwork, see, for example, Sanghera and Thapar-Björkert, 2008; Higate and Sanghera, 2009.

environmental disaster, inequality and political polarization. But the liminal present also contained potentials for a more positive future, not least relating to the environment and racial and economic equality. How do we make sense of the shift of perspective opened up by the pandemic and make use of it as a guide for navigating the future? Einstein purportedly stated that you can never solve a problem with the same mindset or way of thinking that created it.[6]

Repositioning the apparatus of observation is one way to step out of one's 'normal' positionality in order to see the world from a new angle. The pandemic has already paved the way. The challenge is to resist the pull back to old habits that need to change if the planet is to survive and prosper in the future. But what does it mean to change position in space and time, even if only for a momentary glimpse of something different? If the world simply is 'as it is', in all its wondrous materiality, positionality doesn't matter. Truth or falsehood is accessible to me wherever I stand. If my positionality is my 'essence', then it also can't be changed, as it is a given feature of the person and place to which I was born. I am who I am at twenty as at fifty, just as truth and the world are the same wherever and whenever they are observed. Neither orientation takes us further in a situation where everything has been shaken up, where nothing is as it seems.

Observation from a different angle and with a different apparatus can be a strategy to further conversation around deeply embedded habits and assumptions. The obvious problem is how to make this shift if positionality comes with blind spots that shape what is observed. My positionality is a function of experience that began with birth in the US followed by a later repositioning in the Netherlands and United Kingdom. I am academically positioned within the social sciences, which, as Wendt (2015) argues, are dominated by the materialist assumptions of classical physics. My own academic history has been heavily influenced by a philosophical apparatus informed by the later Wittgenstein which resonates with both quantum physics (see Stenholme, 2011; Wendt, 2015) and Eastern philosophy (for example Anderson, 1985). The resonance eases the shift. While a Wittgensteinian or other non-Newtonian Western tradition might be adopted as the apparatus, these would not entirely address the dual concerns of this project, that is, the implications of employing a quantum apparatus in the observation of social phenomena and the potential contribution of thought systems with origins beyond the West. The parallel between quantum physics and the ancient Asian philosophies bridges these two concerns.

---

6 For the discussion surrounding the exact wording of this quote or where it was articulated, see https://en.wikiquote.org/wiki/Albert_Einstein

In a quantum world the apparatus and the position of the observer impact on what is seen. Position is not static but continuously changing and in motion. Roger Ames (1986) argued that there is value in adopting a totally different tradition than one's own for developing a unique hermeneutic perspective. The strategy to be employed here is one of experimenting with an apparatus that is *not* a part of my inherited traditions as a tool for reflecting back critically on the global seats of power from which I emerged. My individual experience or interpretation is not the point, as might be the case for a method of, for instance, auto-ethnography.[7] The objective is to adopt an apparatus that will facilitate an ability to see the blind spots in a context of which I am a part. As such, it is an attempt to see differently than might otherwise be the case, and a choice to investigate the cracks that have been revealed by the pandemic.

Repositioning

While much of the non-Western world finds itself daily engaged with conceptual artefacts with their origins in European and American empire, from the global capitalist economy to the academy, the luxury of choice is often missing. A Western apparatus may be imposed rather than chosen. The experience of adopting a different apparatus is also approximated by anyone who has moved from their place of birth to a different part of the world, even for a short time. One looks back at home differently from abroad. Further, an element of this perspective as strategy can be found in the literature on conflict resolution and the idea that conflicting parties should attempt to place themselves in the shoes of the 'other'.

My last book, *Political Self-Sacrifice* (2012), employed a research strategy that moved in this direction. One of the chapters explored Thích Quảng Đức's act of self-immolation in Vietnam, including the meaning of the act in Mahayana Buddhism. I was fascinated by the association of the act with compassion, in contrast to the disgust or shock with which it tended to be viewed in the West. Fascination led to curiosity and a desire to look more deeply into Buddhism and further to Daoism. *Political Self-Sacrifice* attempted to ground practice and theory in cultural context. The objective here is different.

As my attention and interests travelled East, I came across frequent references to the parallel between these ancient Asian traditions and quantum

7 As a method, autoethnography highlights the influence of the researcher on the research (see, for instance, Briggs and Bleiker, 2010; Dauphinee, 2010). Murphy (2020: 72) notes the potential contribution that quantum theory might bring to autoethnography in light of criticisms that it lacks intellectual rigour.

physics. Several years earlier, I had heard Alexander Wendt present a paper related to what became his 2015 book, *Quantum Mind and Social Science*. At the time, I couldn't understand what he was getting at. As I began to ask questions about the parallel, I contacted him and, in response, he asked me to read a draft chapter of the book. Having developed some knowledge of Buddhism and Daoism by this time, I was much better able to grasp the significance of quantum theory. If approaching quantum subject matter from the position of Daoism and Buddhism made it easier for me to understand what was at stake, I assumed this might be the case for others as well. In the intervening years, I designed and taught a master's level module on agency and strategy in non-Western thought. I began each semester by asking students why they decided to take the module. The most frequent answer was the absence of attention throughout their university careers to perspectives beyond the West.[8] I was fortunate to receive a fellowship from the Leverhulme Trust to rework the material into a book.

The fellowship began in October 2019 and was followed by a series of world-changing events. December saw the election that gave Boris Johnson the mandate he needed to take the United Kingdom out of the European Union. From my position in Scotland, a series of bad weather events made for very dark, windy and wet times in January and February, with large parts of the UK under water. In March the United Kingdom, like much of the globe, went into lockdown, in which we remained for several months, in response to the rapid spread of COVID-19. In May, George Floyd, a Black man, was murdered under the knee of a White policeman in Minnesota, the place of my birth. The protests that followed spread across the world, raising fundamental questions about the persistence of structures of racial inequality that had been exposed and exacerbated by the pandemic. As the year progressed, and variants of the virus emerged, the extent of global economic inequality became impossible to ignore. Stories of rising stock markets and a property boom were juxtaposed with stories of food banks, hunger and spiralling unemployment, as a well as the hoarding of the eventual vaccines by wealthy countries at the expense of the poor. In these circumstances, I could not help but reflect on the relevance of the subject matter of this book for understanding the world that was rapidly changing around me. While I initially resisted the urge to bring these reflections to the analysis, I became increasingly aware that the reflections themselves were an important part of the story, not only for positioning myself as analyst but also for making sense of the substance and relevance of the more conceptual claims.

---

8 This has begun to change in light of the acceleration of efforts to 'decolonize' the academy (Santos, 2017; Arday and Murza, 2018; Bhambra et al, 2018).

The book project began with the objective to explore questions of agency and strategy from the perspective of Buddhism, Daoism and Hinduism. The objective was important for several reasons. First, the task was paradoxical in light of the frequent association of these traditions, particularly in the West, with mysticism and passivity, that is, an assumption that they take us out of the world rather than situating us as actors within it. Second, as suggested in the previous section, the question provided an entry point for deepening understanding of the potential implications of the quantum shift for the analysis of human and social action in the world. The claim that quantum theory may have little relevance to the social sciences because quantum effects wash out above the subatomic level is, arguably, problematic in light of the holism of quantum theory. As Karen Barad (2007: 85) states:

> Quantum mechanics is not a theory that applies only to small objects; rather quantum mechanics is thought to be the correct theory of nature that applies at all scales. As far as we know, the universe is not broken up into two separate domains (i.e. the microscopic and the macroscopic…).

The universe is not broken up into microscopic and macroscopic. Nonetheless, the jump from saying something about the activity of subatomic particles – the preoccupation of quantum physicists – to saying anything meaningful about sentient life, or the kinds of social and political problems humans presently face, *is* a quantum leap.[9] The parallel helps in this transition by looking to traditions that have for millennia explored relevant questions. According to the German physicist Werner Heisenberg, quantum physics suddenly made much more sense and seemed less crazy to him after conversations about science and Indian philosophy with the celebrated Indian poet Rabindranath Tagore (Capra, 1988: 43). The objective here is not to become a physicist. It is rather to adopt a set of tools more appropriate to the analysis of a human context defined by uncertainty rather than predictability. These ancient traditions have a lot to say about mind, action and strategy in the macroscopic world, and thus may provide insight into the practical significance of a quantum 'reality' for the social sciences.

The exploration of the parallel was riddled with contradictions from the start. But contradiction and paradox have a central place in both quantum physics and the Eastern philosophies, which is itself a reason for unpacking the synergies between them. The idea that a particle and wave are entangled is no less contradictory than the Buddhist claim that the self is both real and

[9] A quantum leap is an abrupt change, sudden increase or dramatic advance. In popular discourse it suggests a very large leap, which differs from its meaning in physics.

non-real or the interpenetration of polar opposites *yin* and *yang* in Daoism. Contradiction cannot be accommodated by a Newtonian lens of binary either/or oppositions, or by the law of non-contradiction, which states that contradictory propositions cannot both be true at the same time. At the level of human life, binary thinking translates into stark dichotomies between, for instance, the human or nature, materiality or consciousness, the body or the mind, reality or language, the natural or the cultural, all of which miss the key point of both quantum theory and ancient philosophies that 'life' is entangled all the way up and all the way down. It is holistic and distinctions are drawn from within. The local manifestation of form emerges from formlessness and is part of the production of life in all of its diversity, including humans, who are a part of nature. From this angle, neither the scientist, the analyst nor other agents stand outside the world as objective observers. They are bound up with it. Just as the scientist in a quantum experiment cannot be detached from the apparatus of measurement, human actors always view the world from a particular perspective and position, which is continuously changing.

In a world that has been turned upside down, this book is attempt to seize an opportunity to engage reflexively with the cracks exposed by the COVID-19 pandemic, to look back at Western structures through a reading that begins with knowledge systems that originated in a different place and time but which share a family resemblance with a particular understanding of science. Some of the themes to be explored have resonance with indigenous traditions as well as some strands of Western social theory and post-humanism in particular. As important as these are, I maintain a focus on Buddhism, Daoism and Hinduism, as this was the point of reference for physicists who identified the parallel to quantum physics.

Snapshots from home

The structure of the book revolves around a series of 'snapshots', which together approximate a multidimensional hologram,[10] albeit imperfectly, as a way of engaging with a non-linear phenomenon. The idea of a snapshot of the same object from different angles is a contemporary variation on the statement from the Hindu *Rig Veda*, the oldest of the Vedas, going back to 1500 BCE, that 'truth is one; people call it by different names'. The phrase

[10] 'Hologram' has its etymological origin in the Greek 'holo' meaning whole and 'gram' to write (Bohm, 1980: 183), which Pan (2020) refers to as the 'writing of the whole into its parts'. His definition emphasizes the relational aspects over the popular usage of 'holograph', as a document written in the hand of the author, or 'hologram', which is a three-dimensional image.

conveys an understanding that there is indeed an ultimate reality which is far larger than any one person, theory or cultural position. The problem is that we are unable, with the limited tools available to us, to capture 'truth' in its entirety. A notion of sameness and what makes any two things or phenomena the same is already problematic. While some notion of reality exists, it is far larger than any one of us can decipher from the position we occupy in local space. The classical answer to this problem has been a competition between theories to establish which is more true. The idea of snapshots suggests that we may find glimpses of truth in any one observation or theory, but these glimpses will always be incomplete.

While a snapshot, like a photograph, is a flat image, the hoped for effect, in juxtaposing a series of snapshots from different angles, is to approximate a non-linear and multidimensional hologram. The central point, explored by the theoretical physicist David Bohm (1980: 224–6), is that each part of a hologram 'contains information about the whole object' and is engaged in the totality of movements of enfolding and unfolding, whereby everything is in flux and a process of becoming. Pan Chengxin (2020) argues that quantum holography highlights the ontological duality of relations and things, and sees relations as constitutive of what entities are or become. Traces of the larger whole can be found inside any one entity, which is always relational.

The metaphor of 'snapshots' is drawn on as a structuring device for purposes of exploring the relationship between universe, world and action within it. The central objective is to investigate questions of mind, action and strategy from several different positions, in order to glean the significance of the parallels to quantum physics. A thread connects the different parts, which move from self and mind, to action and strategy, to structure and time, peeling back the various layers of the problem to identify the family resemblances that run throughout in order to more clearly see that nothing permanent lies behind them. What follows is less a linear argument than the constitution of a multidimensional space that begins with problems raised by Buddhism, rotates to Daoism and Hinduism and circles back to Buddhism, while moving through different dimensions of an assemblage of parts which together constitute the whole of action.

Snapshots from 'home' has multiple meanings in the context of this book. First it refers to Wendt's claim that Newtonian physics does not provide humans with a home in the universe, while quantum physics might. As he (2015: 18) states, 'If materialism must stay then consciousness must go, because like the soul there is no place for it in nature. Unfortunately, given that consciousness is widely seen as essential to the human condition, that means there is no place for us in nature either – that we are not "at home in the universe".' Wendt, in his emphasis on scientific debates, has not gone sufficiently far in showing what this home might look like. As Burgess (2018: 129) notes, Wendt 'navigates around most traces of meaningful human experience or social relations that do not already hold social scientific currency.' The ancient traditions provide a means to explore a question of what it means to live a good life and be at home in the universe.

Second, these snapshots were largely composed from my home during conditions of lockdown, which was a particular perspective in time and space, although one widely shared globally. The silences of lockdown, and the lack of movement, created an unusual space to look at the previously taken-for-granted world from a new angle, and not least the relationship between the social constructions of our own making and nature. Lockdown meant a cessation of 'normal' routines and patterns of engagement. With the silencing of the continuous drone of automobiles or airplanes, nature became visible in a way that is not usually the case.

Third, while experiences of lockdown were varied, depending on one's circumstances and position, the pandemic forced a different relationship between individuals and their social world, not least because we were largely confined to home and because, no matter what one's location, we were all threatened by the same virus. In this respect, snapshots from 'home' points to the challenge posed by the pandemic to expand our understanding of home to incorporate the world that we all share, including nature, even while observing and experiencing it from different positions. It returns to the point, compatible with both quantum physics and the Eastern philosophies, that while this whole is always observed from a particular angle, the whole interpenetrates the parts. The cracks exposed by the pandemic revealed that the experience of home was not uniform. Many of those who lived alone suffered from isolation; those who lived with partners and families faced problems of too much togetherness, which in some cases fed abuse. For the wealthy, home became a place of work and school; those who worked for 'essential services', some at the bottom of the 'normal' economic ladder, had no choice but to go to work, to keep society running and to put food on the table, while risking their own lives in the process. Similar inequalities were evident across the globe, if not always in the same form.

The snapshots that follow are not an attempt to determine whether quantum theory or Buddhism is more true, or whether Buddhism, Daoism or Hinduism is more true, or which Buddhism is more true. The objective is rather to identify several points of intersection between quantum science and the ancient Asian philosophies. These points are then used to cut into a 'reality' that differs from the Newtonian and ask what this might offer, not only for the study of the social and political world but also for navigating the potentials that might emerge from the devastation of the pandemic. As Karen Barad (2007: 148) states, 'Apparatuses enact agential cuts that produce determinate boundaries and properties of "entities within phenomenon".' The apparatus employed here arises from curiosity about a parallel that has frequently been articulated by quantum physicists, as well as Buddhists, and not least by the Dalai Lama (2005).

I am not a quantum physicist, and I assume that most readers will have little or no background in physics. Most social scientists, including those who operate on the basis of more materialist assumptions derived from classical physics, are also in this category. I have never read an article or book by a social scientist that began by explaining the assumptions of Newtonian physics before proceeding with a materialist analysis. They don't need to because they build on accepted wisdom. By contrast, those who wish to explore quantum theory are continuously confronted with questions of what it has to do with the reality that social scientists study (Murphy, 2020: 37). What follows is an attempt to explore how quantum dynamics reveal themselves in human social and political practice in contexts of uncertainty. Quantum impermanence, complementarity and entanglement and their significance for understanding mind, action and strategy are discussed in the next chapter and threaded throughout the book, but debates about how quantum physics has evolved and the different schools of thought within it do not form its substance. Any one snapshot is multidimensional in its exploration of an aspect of the parallel, its relevance for understanding a particular context of practice and its potential implications for how more contemporary global processes are understood. Each snapshot zooms in on a particular angle of action in a different space and time.

Theories of IR or analyses of empirical contexts are often criticized for providing *only* a snapshot without adequate attention to context or history. While reinforcing the importance of taking both context and history seriously, the central claim here is that snapshots are all that we can hope for. The claim has implications for what we assume about ourselves as analysts as well as about the world of analysis. IR theories often imply that they capture reality 'as it is', or an approximation of it, when it would be better to view, for instance, realism or liberalism as themselves snapshots that capture an element of international life but by definition only a partial one, which is conventional in its construction. Whether the agents are states or rational economic actors, both are products of human design rather than nature. The theories provide a flat picture of a world defined by the primacy of states or the primacy of markets, without consideration of where these institutions come from or whether they represent the only potential for constituting 'reality'.

A hologram, by contrast, is a multidimensional image wherein everything is infused with everything else and the contours of each pixel are reproduced in the whole. It may be impossible to capture the universe in its totality, but we gain a deeper understanding of our place within it by observing the 'same' phenomenon from different angles. If everything is part of everything else, then, as L.H.M. Ling (2013) argued, East is in West and West is in East.

From this angle, it is difficult to think of cultures as contained phenomena which possess ownership of certain ideas or practices. This is not to deny that ideas and practices take root and flourish in a time and place and that particular forms of ritual and practice are more at home in some cultures than others. It is rather to emphasize Wittgenstein's (1958: para. 67) point that we are looking for family resemblance rather than essence.

The intention is not to deny the long and troubled history of appropriating ideas and practices from one part of the world and owning them in another, which takes many forms. It is rather to view the world as a richly diverse tapestry, woven through millennia of practice, wherein threats and threads starting in one space have travelled to others. The concern is not to identify which version of quantum theory or Buddhist or Daoist practice is more true, but rather to explore what these ancient philosophies might tell us about the macroscopic implications of quantum physics for understanding strategic potentials for navigating an uncertain world. The issue is not science (reality) vs. culture (interpretation) but rather science and culture.

The incongruity between Newtonian and quantum metaphysics also arguably rests on a cultural contrast. While both constructs are products of the West, the quantum more closely approximates discoveries that informed the observations of ancient cultures. The Newtonian worldview assumes determinism, mechanism, locality and an emphasis on brute matter. The quantum is about indeterminism, non-locality, fluidity and the entanglement of matter and consciousness. The contrast between two sciences becomes cultural when placed in historical context. The structures of modernity and the Enlightenment, now just a few centuries old, are heavily indebted to the mechanical view of reality provided by Sir Isaac Newton, among others. As Wendt (2015: 12) argues, most social scientists, including constructivists, are indebted to this tradition of science.

Practices of Western colonialism and imperialism were also arguably indebted to this tradition of science, as were traditions of modern thought from Hobbes and Locke to more contemporary liberal and realist thinkers. Particularly since the beginning of the nineteenth century, the globe has been carved up into states, surrounded by boundaries, which came to be understood as contained cultural units, even while these boundaries in practice cut across and through extant cultural loyalties. Quantum physics, while also originating in Europe and the US over the last century, articulates a much different view of reality, but one that is in sync with the assumptions of ancient philosophies developed millennia ago across Asia. Ironically, these same societies had, in modernity, been silenced by practices of Western intervention and dominance, not least because they were said to lack civilization and science.

Snapshots from Home points to the profound implications of grounding reality in one science or the other for how we understand life and our place within it. As already suggested, the contrast raises a question of whether indeed we can ever be at home within a purely Newtonian universe. That which constitutes life, and not least consciousness as opposed to pure matter, is absent in a purely Newtonian world. Speech, emotions, the breath of life, perception and consciousness have no place in a world of mindless matter. The classical scientist, while far from mindless, is first and foremost rational (but not emotional) and concerned with physical measurement (but not consciousness and intention), and treats language as epiphenomenal (as representing a priori objects) rather than as a means by which communication takes place within a sentient and ever-changing relationality. The observer stands outside the object of observation.

While caricatures, these oppositions inform not only life in the laboratory but a way of life that has become increasingly global. In the classical view, we are what we are in the physical reality of our separate bodies, which is more about material existence than consciousness. Separateness is prior to any relationship to other life, human and otherwise. A relationship is thus less an emotional bond than a rational calculation that maintains separateness in the face of onslaughts of various kinds, which in the extreme form constitutes a Hobbesian 'state of nature'. Our grounding in this world of separateness and separation has, at least since the Enlightenment, made us strangers to one another and all other life. The structural qualities of the world do not express the value of human life and bonds but rather of the trade in objects for the profit of some at the expense of others.

Home and family are relational phenomena which look different across cultures. To contemplate what being at home in the universe could potentially mean is a larger conversation in itself. But some notion of home and family that resonates across cultures necessarily takes us beyond the contained nuclear family of post-World War-II life in the West. The danger in suggesting that quantum physics, or the parallel to Eastern wisdom, provides a *better* point of departure for imagining 'home' is the idealization of an image that is not or cannot be borne out in reality. As will be discussed in the next chapter, some physicists have ridiculed the parallel as pointing to an earthly paradise. Others might point to the use of quantum mechanics in the development of the most extreme form of destruction, as manifest in atomic weapons. Equally, for better or worse, globalization has arguably already propelled us down the path to a new home, one characterized by non-locality and entanglement. This new home is situated in an emerging global technological culture and interconnections formed through social media and the World Wide Web. The different potentials hinge on the extent to which limits are imposed by causal closure. Does the material world impose limits that constrain options, requiring a 'realistic' view of what is possible given the world 'as it is'? Or are we, as suggested by the American theoretical physicist John Wheeler, participants in creating the world? Is our seeing, thinking or acting in the navigation of 'reality' merely a reaction to a pre-existing world, or constitutive of it? The answer is not either/or but both. The series of crises in 2020 have turned the world upside down. The degree of uncertainty and change has disturbed any notion of the world 'as it is' and opened space for reflecting on our ability to shape future potentials.

Snapshots from different angles

Contrary to Newtonian assumptions that the world is composed of mindless matter, Buddhism, Daoism and Hinduism, like quantum physics, see a universe of continuous movement, of flux and change, propelled by a dynamic relationship between matter and consciousness. The source of this dynamism is the mutual implication of particle and wave, or what is referred to in quantum physics as their complementarity. If in classical physics a particle is a particle and a wave is a wave, in quantum physics a particle can, in certain circumstances, become a wave and a wave can become a particle. The mutual implication of particle and wave turns the Newtonian conception of 'brute matter' on its head. The very thing that realists point to as 'reality', which our language is said to merely describe, correctly or incorrectly, is formed through language and is thus dependent on context and the manifestation of potentials. In Daoism, Buddhism or Hinduism, this suggests two realities or truths, although they are mutually implicated. The one is more universal, relating to a consciousness and intelligence that is entangled and infuses all material life. The other is more conventional, resting on an illusion of separateness, and, at the level of humans, constituted in language. While 'Truth is one', according to a famous verse of the Hindu *Rig Veda*, truth can have various names. The claim is less a celebration of relativism than a recognition of the influence of one's position on what is seen and a statement that any one object of observation has many different angles from which it can be viewed.

The idea of snapshots of the 'same' object from different angles underlies the structure of this book. The main concern is to examine conceptions of mind, action and strategy in Buddhism, Daoism and Hinduism in parallel to quantum impermanence, complementarity and entanglement. Each of these traditions contains a paradox regarding the relevant concepts, but the focus of each is somewhat different. Buddhism highlights the paradox of self that lacks an intrinsic nature, and a position of 'no mind' from which action arises. Daoism, which contains a similar understanding of the self, highlights a notion of actionless action, or *Wuwei*, and *yinyang* strategy. Hinduism places emphasis on karma, which means action, and the fruits of action, both of which are closely linked to a concept of time. The concept of karma is shared by Buddhism but takes on a somewhat different meaning. These are snapshots that together provide a fuller picture of the agentless agent who engages in actionless action as she strategically navigates the universe, her selfishness tempered by the entanglement of all life. The process of moving through the snapshots, can be thought of in terms of a further metaphor, more at home within Daoism: the process is akin to walking around a mountain and watching the light shift along with the movement and the changing geography, weather or time of day. The mountain represents a whole, which is at one and the same time part of a larger whole, a mountain range; what is seen by the observer, as her location in time and space changes, is always partial. The shifting apparatus moves toward a more complete view, even while its capture is beyond the realm of the possible.

The focus of each snapshot is a concept relating to different parts of a composite assemblage of action. I ground the concepts in philosophies that parallel quantum physics as well as examples of practice that illustrate the working of action in the world. As suggested by Figure 1, each reconfiguration of the overlapping lenses makes it possible to zoom in on a particular aspect of mind, action or strategy from a particular angle.

The tasks represent a particular cut into complex and diverse materials, guided by an apparatus that is positioned at the intersection between quantum science and ancient Eastern philosophy. The non-linear presentation might be thought of as what Emilian Kavalski (2017: 4–5) refers to as 'itinerant translation', which 'draws on a tradition of intellectual wanderings, peregrination, and trespassing, which does not subscribe to linear logics of detachment, coherence and parsimony … a narrative journey of fusion, exchange, dialogue and contestation.'

Figure 1: Overlapping lenses

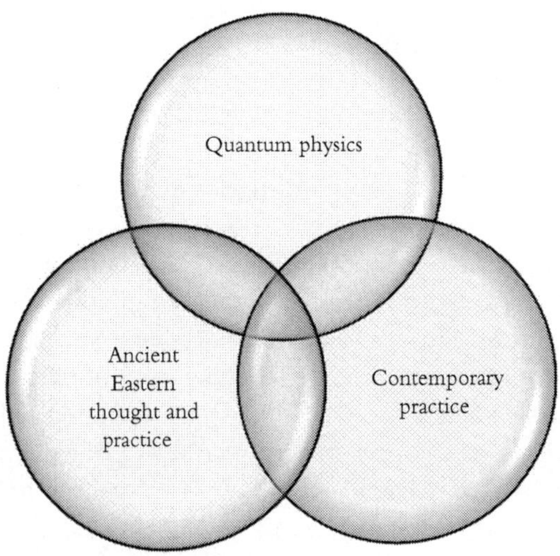

The context of the 2020 pandemic was marked by an imperative to 'listen to the science', against a backdrop of a politics in which 'fake news' had a significant place. The intention here is not to denigrate science but to question the assumption that it is unitary, while situating Newtonian science within a quantum world. An emphasis on the position of the observer should *not* be understood to endorse fake news. Instead, the objective is to highlight the conventional nature of the apparatuses by which truth is negotiated and the importance of conversation for navigating multiple crises that are global in nature.

In what follows, each angle represents an attempt to reconfigure the relationship between formlessness and emergent form, the unseen and the seen, potential and power to assess what they might have to say about navigating an uncertain future. The second part of the introduction explores the parallel, why the parallel is important for conceptualizing mind, action and strategy, and its significance for how the relationship between science and culture is understood.

The three sections that follow each contain two snapshots which zoom in on the same parallel from a slightly different angle of action. Section one explores the consciousness/matter relationship from the perspective of quantum impermanence as it relates to Buddhist concepts of self and mind. Snapshot 1 introduces the reader gradually to the more abstract Buddhist concepts of self. The 'Mindfulness Revolution' announced by *Time* magazine in 2014 provides a foil for exploring the illusory sense of 'I' within Buddhism and the philosophy of emptiness and dependent origination articulated by the second-century Buddhist philosopher Nāgārjuna. Snapshot 2 moves to an exploration of the relationship between emptiness and 'no mind'. Mindfulness is usually associated with some form of meditation. The latter is also central to Eastern traditions of fighting. Building on an example from Japanese Zen Buddhism, which is a strand of Buddhism influenced by Daoism, the snapshot explores three interrelated paradoxes relating to mind, action and ethics. Returning to the controversy around mindfulness in the West, the snapshot concludes with a reflection on the ethical potentials inherent in using mindfulness techniques, among other things, for the treatment of US soldiers who suffer from post-traumatic stress disorder (PTSD) as a result of their experiences in Afghanistan and Iraq.

Section II continues the shift to Daoism, providing snapshots that reveal the dynamics of quantum complementarity and Daoist yinyang. Both snapshots highlight the importance of contradiction or paradox for understanding the dynamics of action and strategy. Snapshot 3 examines the Daoist concept of actionless action (Wuwei) and its significance for thinking about the environment. The climate change debate has been framed as a conflict between those who deny the science and activists who encourage leaders to look to the science. While the latter confirms that climate change is indeed happening, many of the developments that have led to the current crisis also have their origins in science and the industrial revolution it produced, as well as in assumptions that humans can control and should control the environment. How does the relationship between the human, nature and environment change within a Daoist framework? What is the meaning of actionless action or Wuwei, which itself rests on a paradox, within this understanding of nature? What might actionless action mean for how we deal with the threat of climate change? Snapshot 4, staying within Daoism, then moves to an exploration of Sun Tzu's *Art of War* and yinyang strategy. Strategy, in this conception, does not provide abstract blueprints for achieving objectives. It instead involves the navigation of the potentials of a context. The war on COVID-19 is approached from the perspective of Sun Tzu to see what difference this makes in the response to the global pandemic.

Section III shifts to a further parallel between quantum entanglement and karma, both of which point to action and relationality that extends across space and time. Snapshot 5 explores the concept of karma as it relates to the consequences or fruits of action, and zooms in on Gandhi's experiment in nonviolence. Gandhi, as a strategist, was, like the quantum scientist, bound up with the apparatus of measurement. His experiment provides an illustration of the quantum notion, articulated by John Wheeler, that we are participants in creating the world. The example reveals what is at stake in the shift from the separateness of bodies to an understanding of life as constituted in relational entanglements. Snapshot 6 realigns the apparatus slightly to the Buddhist concept of karma. From a Buddhist perspective on karma as habit, it is less the transmigration of an individual soul than the relational properties of a culture that are transferred from one generation to the next. Karmic entanglement is used as a tool to explore the politico-ethical implications of the distinction between things and relations as it relates to the defunct legal category of the chattel slave, or the human as property. The analysis starts against the backdrop of the murder of a Black man by a White police officer and the subsequent spread of Black Lives Matter protests across the globe. It explores the reproduction of karmic residues, rooted in the transatlantic slave trade and chattel slavery, which arise from an entangled system defined by a fault line distinguishing Black from White. Some of the snapshots provide insight into the working of the parallel in a culture of origin; others provide an apparatus for 'seeing' an aspect of Western experience and culture from a different angle.

The three points of parallel – impermanence, complementarity and entanglement – give insight into the problem of human suffering, the significance of contradiction and paradox for deepening our understanding of the dynamics of action and strategy, and the reverberations of global entanglements across space and time. The six snapshots begin from and return to Buddhism, which represents less a beginning and conclusion than one point of intervention among many within the hologram. The circle starts and ends with a question about the underlying assumptions that inform these traditions and what happens when they are transported to an individualist, capitalist culture. This circular journey provides snapshots from different angles of mind, action and strategy in an uncertain quantum world.

The conclusions seek to develop the implications of these snapshots for thinking about what it means to have a home in the universe, and not least how this relates to re-envisioning life after pandemic. What have we learned? If humans are not at home in the universe of Newtonian science, how does this hermeneutic strategy help in constructing the contours of a different conversation about what our global home might look like? How does an appreciation of consciousness reconnect humanity with life more broadly and open possibilities for negotiating the world and the universe without doing harm? The final chapter examines the implications of the analysis for how we navigate further changes in a post-pandemic world.

The parallel between quantum physics and ancient philosophical traditions reshapes the apparatus of navigation and transforms the agent. It provides an alternative hermeneutic sensibility for acting in a world where we are understood to be participants in creation. The language of 'we' here requires some specification. The social sciences, and IR in particular, have grown up in one part of the world, that is, the West, centred in empires dominated by the United Kingdom and the United States. In this world, we employ a scientific apparatus which stands outside the world, seeking 'truth' about its nature. The pandemic has opened up an opportunity to reconfigure the 'we' and to view the present construction of the world from a different angle. While the focus on the ancient Asian philosophies may look like yet another cultural appropriation, all too characteristic of an imperial past, it is not about the practices of any one religious system informed by particular rituals and practices. It does not identify with any existing form of government or governance, whether Asian or otherwise, nor is the critique purely of the West. Indeed, capitalism is as pervasive in communist China as throughout the rest of the world. Rather, the Eastern philosophical systems are prior to any particular cultural or religious manifestation. The intention is not to reduce the West to a monologue of possessive individualism, or to deny the diverse traditions of thought and practice that have emerged from this corner of the world. It is instead to explore some of the cracks that emerged during the pandemic from an angle other than 'normal' for what they reveal about habits of modern life that we would normally avoid looking at or seeing. If we are going to survive some of the global challenges that lie ahead, a more diverse and global conversation (Fierke and Jabri, 2019), resting on mutual deference and respect, is needed.

The positive intention is to reimagine a shared home in the universe within a system that conceives of home as more than the rational construction of a building with foundations but which expresses a potential for emotional warmth and consciousness as well. Homes are rarely utopian but they do constitute a space where all life matters, where the participants are conscious of their relationship to others and where these interconnections are devoid neither of rationality nor of deep emotional connections. Families that stay together find ways to navigate their emotional relationship without doing damage to the integrity of the whole. There may be a temptation to think about home or family as a level of analysis, which can be separated from other levels above, that is, governments, or below, for instance servants who prepare the food and animals who provide it. The key point is to contribute to a conversation about the nature of a quantum family that is holographic, where the relationships contained within any one pixel are reproduced throughout and across space and time, even while they are continuously changing. The task may seem utopian at a time when politics across the world have been marked by deepening divisions of hate, but it is arguably a necessary condition for facing and surviving existential threats that are global in nature, as has become so painfully evident in the context of a global pandemic. Reimagining home in a rich tapestry that is entangled is about the survival of life. Such reimagining requires that we ask not only what it means to survive but also what kind of life we want to live.

Getting to Know the Apparatus

Repositioning the apparatus impacts on what is seen. The rationale for repositioning and the implications for my positionality as researcher were explored in the first part of the introduction. Here we seek to deepen understanding of the apparatus, that is, the parallel between quantum physics and ancient Asian philosophies, as preparation for the snapshots that follow. Nearly a half century ago Fritjof Capra's ([1975] 1991) *The Dao of Physics* drew a parallel between quantum physics and Eastern mysticism, focusing in particular on Buddhism, Daoism and Hinduism. His book, which was a sustained exploration of the parallel, was both insightful and popular. The argument resonated with the New Age politics of the 1970s. In the context of the argument that follows, Capra's framing of quantum science and Eastern mysticism is problematic in so far as experience within the world is the primary concern here. I argue that an earlier articulation of the parallel, by the Danish physicist and father of the Copenhagen interpretation Niels Bohr (2010), provides a better positioning for exploring the meaning of mind, action and strategy in an uncertain world. Bohr argued that mysticism was not the point of the parallel, and shifted emphasis to our role as spectators and actors. Further, Bohr's concept of complementarity provides a different angle on the relationship between science and mysticism that goes beyond the either/or terms in which it is usually discussed.

The first part of this chapter examines various ways that physicists have articulated the parallel between quantum physics and the Eastern philosophies, elaborating in particular on the contrast between Capra and Bohr. The second part examines the contradiction of real and non-real and unpacks three specific parallels, including impermanence, complementarity and entanglement, as they relate to mind, action and strategy, which provides the apparatus for making 'cuts' – to use Barad's terminology – into the various contexts of practice observed in the snapshots. Part three explores how an emphasis on entanglement in language shapes the relationship between the natural and social sciences. The fourth part then revisits the relationship between science and mysticism, demonstrating a conception of the two as intertwined rather than mutually exclusive. The final section returns to the tendency to identify Buddhism, Daoism and Hinduism, like other non-Western traditions, with culture and religion as distinct from science.

While each has a religious expression, and Hinduism is a religion, these traditions originate with philosophies that resonate with the discoveries of quantum physics.

The parallel

Popular representations of quantum physics have alternatively embraced or created distance from its parallel to Eastern wisdom. Quantum mysticism is considered by many scientists to be a form of quackery (Grim, 1982; Stenger, 1997; Shermer, 2005; Collins, 2010; Pigliucci, 2010). Karen Barad (2007: 67–8) criticizes those who 'often position [quantum physics] as the scientific path leading the West to the metaphysical Edenic garden of Eastern mysticism'. She further states that 'those who naively embrace quantum physics as some exotic other that will save our weary Western souls forget too quickly that quantum physics underlies the workings of the A-Bomb'. Yet, Robert Oppenheimer, who is seen as the father of the atomic bomb (Temperton, 2017), responded to his observation of the first atomic test with a passage from the Hindu *Bhagavad Gita*: 'If the radiance of a thousand suns were to burst into the sky, that would be like the splendour of the Mighty One and I am become Death the destroyer of worlds.'

Quantum mechanics gave rise to the atomic bomb. An atomic physicist gave meaning to a moment of awe by reference to one of the most ancient and revered texts. Oppenheimer (1954: 8–9) further stated that 'what we shall find [in modern physics] is an exemplification, an encouragement, and a refinement of an old wisdom [in Buddhist and Hindu thought]'.

There is a mystical element to the Asian traditions which I do not wish to discount. Indeed, science and mysticism have also combined in the West. The Freemasons, a seventeenth-century fraternal order including Sir Isaac Newton, sought to combine alchemy with science (Bauer 2007), an observation which becomes significant later in the argument. The point here is that the awe inspired by the explosion of the first atomic device suggests that the problem is more complex than an either/or choice between mystical bliss and nuclear destruction. Capra explores a range of very interesting and relevant parallels. However, his association of science with the West and religious practice with the East, and his contrast between rational scientific inquiry and the intuitive pursuit of the mystic, detracts from recognition that Buddhism, Daoism and Hinduism also gave rise to systematic forms of inquiry.

Hinduism, going back to the first millennium BC, developed methods of a 'supreme science' (*Brahmavidya*), which was concerned less with the external world than with knowledge of the reality that might be discovered through the mind. Similar practices were undertaken by practitioners within the *Madhyamaka* tradition of Buddhism, combining both reason and meditative investigation (Dalai Lama, 2005). The Eastern scientists did not conduct experiments that fired tiny quantum particles at multiple slots, nor did they build expensive weapons or equipment. They instead used meditation techniques to explore the nature of reality within their own minds (Smetham, 2010: 8–9). Western science highlights the third-party observation of the universe by the scientist, while the focus of the ancient science was the first-party experience and observation of the meditator. As forms of science, neither would have much significance for a 'social' science or forms of observation relating to the everyday world. However, as the snapshots demonstrate, the ancient science of mind had more practical implications and applications as well.

Capra's framing is conducive to a separation of practices of meditation from action within the world.[1] The importance of the latter is highlighted by recent discoveries that suggest the mutual influence of the *Daodejing*, a focal point of Daoism, and Sun Tzu's *Art of War*, one of the greatest works of strategy the world has known (Li, 2000; Ho, 2002; Yuen, 2014). As Yuen (2014: 6) states, 'It has been long believed that the *Art of War* and Taoism are interlinked ... where Lao Tzu (the person) appears to have had an impact on Sun Tzu's thought, it also appears that *Tao Te Ching* may have borrowed extensively from the *Art of War*, as the latter preceded *Tao Te Ching* chronologically.'[2] Sun Tzu's work has been adopted by many across the world, often without due regard for its entanglement with Daoism. Analyzing the working of Daoism within Sun Tzu's strategy, and vice a versa, reveals the inseparability of effective action within the world and an entangled web that extends from the mind and body through various layers of context to the universe.

What the Eastern philosophies show us, which physics on its own cannot, is a non-dualistic understanding of the relationship between day-to-day movements within the world and a conception of the universe. From this angle, the subject matter of physics and that of the social sciences are entangled rather than existing in a hierarchy of difference where the latter is merely the poor cousin of the former. In a quantum world, where nothing exists independently of anything else, and where parts are holographically reflected in wholes all the way down and all the way up, a dividing line that separates the microscopic from the macroscopic, as is frequently articulated within the quantum social science debate, is misleading. If subatomic motion, arising from the dynamic relationship between particle

[1] Although a later book (Capra, 1997) does explore the relevance for more socio-political phenomenon, and for deep ecology in particular.

[2] I have maintained the author's spelling in this quote but use *Daodejing* in the context of the larger book. Throughout the book I rely on the Chinese pinyin in providing phonetic transcriptions. The exceptions are Lao Tzu and Sun Tzu as these names, based on the Wade-Giles romanization, are so familiar in a Western context. The Chinese pinyin would be Laozi and Sunzi.

and wave, is characteristic of all life, then questions can be raised about the assumption that activity at one level is isolated from activity at another, or indeed whether the physicist's role in observing subatomic particles might be understood as complementary to the task of the social analyst rather than as mutually exclusive. That early quantum theorists were themselves impacted by Eastern philosophy suggests a much different relationship. The social science debate revolves around questions of how to move from the microscopic to the macroscopic, or whether quantum social science is based on analogy or actuality (Murphy, 2020: 44). The physicist Werner Heisenberg, by contrast, started with a human form of life as a backdrop for understanding physics. Engagement with Indian philosophy both enhanced his understanding of quantum physics and led to a conclusion that the latter could not be all that crazy if an entire culture had subscribed to similar ideas (Smetham, 2010: 43).

A further articulation of the parallel between quantum physics and Eastern philosophies is a useful starting point for exploring its implications for mind, action and strategy in the world. In a 1937 lecture in Italy, the Danish physicist Niels Bohr (2010: 20) stated:

> For a parallel to the lesson of atomic theory … we must in fact turn to … [the] kind of epistemological problems with which already thinkers like Buddha and Lao Tse have been confronted, when trying to harmonize our position as spectators and actors in the great drama of existence.

Bohr's parallel regards 'our position as spectators and actors'. In the context of a debate with Einstein, Bohr (2010: 20) argued that mysticism is not the point of the parallel. He further stated that it does not imply the acceptance of any 'mysticism foreign to the true spirit of science' (2010: 20). The parallel instead provides a point of departure for examining whether 'the straightforward solution of the unexpected paradoxes met with in the application of our simplest concepts to atomic phenomena might not help us to clarify conceptual difficulties in other domains of experience' (Bohr, 2010: 20). The domain of experience that is of interest here is mind, action and strategy at the macroscopic level. Far from suggesting an earthly paradise, or mere quackery, Buddhism, Daoism and Hinduism relate to millennia-long traditions of systematic inquiry into human suffering and the relationship between consciousness, materiality and ethics. While Hinduism is a religion, and Buddhism and Daoism have religious expressions which came later (King, 2015), the purpose of this book is to explore the conceptions of mind, action and strategy contained within these philosophies and the implications of these conceptions for understanding life in an uncertain world.

Quantum physics has led to developments that have revolutionized human life, from the atomic bomb to television, scanners, computers, iPods, MRI machines, mobile phones or skype, to name just a few. While we take these technologies for granted, there remains a reluctance to look more closely at the implications of a quantum turn for life more broadly. There is reason, and not least the widespread experience of uncertainty during the pandemic, to take the parallel seriously and to assume that it does *not* rest on 'naïve romanticization' (Barad, 2007: 68). Bohr's parallel poses the question of why he drew it and the implications for a quantum world above the subatomic level. The ancient traditions demonstrate the significance of impermanence, complementarity and entanglement for conceptualizing human mind, action and strategy.

Real and non-real

Bohr's articulation of the parallel refers to the kinds of epistemological problems confronted by Buddha and Lao Tzu. The problems arise from assumptions embedded in ordinary language that make a clear separation between the behaviour of objects and the means of observation. These assumptions underpin classical physics, even while they are not entirely justified by all everyday experience (Bohr, 2010: 25). Bohr himself is said to have provoked an 'epistemological earthquake' (Omnes, 1999: xxi) regarding the role of the observer and the apparatus of observation in what is measured and thus known. While classical physics assumes an observer who is independent of the world to be measured, Bohr's discovery highlights the interaction between ourselves and the world, stressing that humans intervene at every stage of science and not least in determining the very language in which laws of nature are formulated.[3] The observer is entangled in language, and the scientist, in talking about phenomena that do not follow the laws of classical physics, is, nonetheless, dependent on more classical concepts. As Aage Petersen (1963: 11) noted:

> When one said to [Bohr] that it cannot be language which is fundamental, but that it must be reality which, so to speak, lies beneath language and of which language is a picture, he would reply 'We are suspended in language in such a way that we cannot say what is up

[3] Bohr's shift does not require abandoning the rationality and objectivity which is central to scientific practice but rather suggests the possibility of conveying the meaning of phenomena in such a way as to provide 'equivalent information to all observers', or statements that are intelligible to all human beings (Rosenfeld, 1961: 384).

and what is down. The word "reality" is a word, a word which we must learn to use correctly'.

Reality does not lie beneath language. Our words do not mirror an independently existing world. The meaning of the word 'reality' thus changes and we must learn to use it correctly. Reality rests on paradox.

The frequent assumption that quantum effects wash out above the microscopic level misses the fundamental point that, in a quantum world, the apparatus and the observer are entangled, which makes the key issue one of *meaning* and how knowledge is generated. Mathematics is the language used in quantum mechanics to say something about the nature of reality, and indeed whether any reality exists beyond the mathematics is a point of debate.[4] In reference to more contemporary debates, Philip Ball (2018: 31) claims that meaning and the generation of knowledge *is* the key point, rather than the interaction of wave and particle, which can be an impediment to our understanding. Bohr argues that actors of all kinds are entangled with the apparatus of observation, that is, particular meaning structures that shape the reality that is seen. The Dalai Lama (2005: 74) notes that the central philosophical problem posed by quantum mechanics is whether 'the very notion of reality – defined in terms of essential real constituents of matter – is tenable.' The concept of emptiness in Buddhist philosophy, he claims, provides a coherent model for understanding a reality that is non-essentialist. Quantum science and Buddhism contain parallel constructions of 'reality' as both real and non-real at the level of philosophy; both raise questions of meaning, but the practical implications of the two approaches are different. Science seeks to establish how the world and universe work, and is focused less on ethics than technological potentials. Buddhism is concerned with the transformation of human suffering, which is an ethical concern relating to human potentials. The three sub-sections that follow explore a series of contrasts between Newtonian and quantum mechanics and specific parallels between the latter and Buddhism, Daoism and Hinduism. The objective is to provide the broad outlines of three points of parallel rather than a detailed analysis of the intricacies of either half. Each parallel should be viewed as a guide to understanding the working of the apparatus within the snapshots, which will further unpack the assumptions and concepts surrounding mind, action and strategy.

4 Both quantum physics and quantum mechanics study subatomic particles. Quantum mechanics is more specific in referring to the field once it is formulated into mathematical laws. I have avoided the term quantum mechanics, except where directly relevant, to highlight the point that analysis of more macroscopic phenomena relates to languages other than mathematics.

Mind

Both quantum mechanics and the Eastern philosophies express a relationship between real and non-real. Both turn the Newtonian assumption that reality is composed of mindless matter on its head. Newtonian physics highlights 'things', which are assumed to have ontological properties that exist independently of the observer. A particle is a particle and a wave is a wave. The seminal shift that gave rise to quantum physics emerged from the famous double slit experiment and the observation that a particle can become a wave and a wave can become a particle in certain circumstances, or what is referred to as wave-particle duality.[5] The experiment illustrates that elementary particles are not objective material objects with characteristics that can be determined. Rather, phenomena arise from an interaction of some kind, or indeed can be seen as the interaction itself. In classical physics, a particle has a path, position and velocity that exist independently. Instead, as Stapp (1971) further stated, a particle is 'a set of relationships that reach toward other things.'

The things of Newtonian physics are material and deterministic. The relations that reach toward things in quantum physics arise from a wave-particle duality that is indeterminate at the microscopic level. Buddhism provides an account of how this indeterminism, arising from a duality of real and non-real, relates to the impermanence of life and human suffering. For both quantum physics and Buddhism, the apparent substance or essence of matter – or of the individual self in Buddhism – is an illusion. Non-real in this case does not mean non-existent, but rather points to the absence of essence. Reality is to be found in the *relational* constitution of 'things' or humans and other life. Reality is fundamentally relational in so far as 'things' or 'selves' are shaped through relationships, causes and conditions that are continuously evolving.

---

[5] The experiment was first conducted in 1801, long before quantum physics, by Thomas Young, and has been reproduced numerous times in the centuries since with identical results. In the experiment, an apparatus placed at one of end of the room shoots out quantum or elementary particles toward a barrier with two small holes to the end of measuring the way that they are detected after coming through the openings. The expectation would be that particles begin and end as particles. Instead, something strange happens. With a single opening the particle passes through the barrier as a particle, as one might expect. In the presence of two slits the same electron simultaneously passes through both slits, as only a wave of energy can do. Thus, when there are two openings, the particle acts like a wave, moving through two slits at once, while arriving at its destination as a particle. The question is how the electron perceives that the second opening exists. Since the electron can't really 'know' anything, it was concluded that the only source of this awareness is the person watching the experiment. Knowledge that the electron has two possible paths to move through must then be in the mind of the observer.

Many quantum scientists have argued that the nature of reality is a problem of mind rather than matter. For instance, Max Planck (1944) argued that 'all matter originates and exists only by virtue of a force … We must assume behind this force the existence of a conscious and intelligent Mind. This Mind is the matrix of all matter.' Wendt's (2015) book is an elaboration of quantum mind and its significance for social science. While noting the powerful grip of materialism on the scientific imagination, for Wendt (2015: 81) the problem, from a purely materialist position, is that one can't get to consciousness, which he defines as the experiential aspect of mind (2015: 15). Quantum physics raises a question of whether the 'ultimate constituents of reality are indeed wholly material', given that physicality, in his argument, is not equivalent to materiality. The 'collapse of wave functions', he argues, provides a physical but non-material space for consciousness at the quantum level (2015: 109). In this respect, 'mind is neither reducible to matter nor emergent from it. Mind is in matter all the way down, which makes mind and matter continuous' (2015: 112).

Buddhism brings the real/non-real duality to the matter/consciousness relationship, while also providing insight into the ethical implications. The first section of snapshots explores a notion of self as both real and non-real and its relationship to mind and consciousness. The second-century philosopher Nāgārjuna's concepts of 'emptiness' and 'dependent origination' illustrate the seamless and mutually inclusive relationship between a universe that is lacking in ontological primitives and forms of life. Because everything depends on a web of interconnectivity, phenomena are empty of essence or 'own being', and, thus, in a manner of speaking, are non-real, even while they are real as relational phenomena. Within the Madhyamaka tradition of Buddhism, emptiness is a creative realm of indeterminacy, beyond language yet inextricably bound up in it.

The duality of human life regards the tension between the experience of 'I', associated with the ego self, and a more relational self. The former, observed through a classical apparatus, contains an essence of separateness from others, which is reinforced by distinctions between 'mine' and 'yours'. Suffering arises from the difficulty of looking at our mortality and dependence on others. The attachment to self-protection may give rise to the poisons of ignorance, hate and greed and further manifestations of harm to others. Repositioning the apparatus involves a rotation away from the ego self toward the relational and entangled self, which opens up potentials for compassion.

All humans are suspended in language, no less than the scientist. The thoughts that run through the mind are expressed in language. The activation of thought arises from streams of consciousness and an activity of 'I'-ing. For both quantum physics and Buddhism, any particular activation is more a function of intention than of free will per se, a distinction that is unpacked in Snapshot 2. An apparatus of mind that observes through narratives of the ego self and its interests acts within one world; an apparatus that is repositioned to observe the entanglement of life and suffering enacts another. The two positions, which represent more a spectrum than an either/or choice, have ethical consequences. Buddhist practices of meditation provide a method that allows for an experience of emptiness that arises from cessation of the proliferation of concepts. The subsequent calming of mind provides a particular perspective from which to re-engage with life and act in new ways. The parallel between quantum and Buddhist notions of real and unreal does not represent an analogy or a translation from microscopic to macroscopic, but instead the application of 'a coherent model' of a non-essential reality to different domains of life and experience.

Action

Bohr's parallel refers not only to the Buddha but also to Lao Tzu and Daoism, which highlights a related but somewhat different issue. Bohr developed a concept of complementarity which has a parallel in the Daoist yinyang. Complementarity and yinyang express a relationship between polar opposites that are at one and the same time mutually implicated and mutually exclusive such that both cannot be seen at the same time. The relationship between opposites is not static but continuously changing. Complementarity concerns a problem of measurement. In a more classical conception of reality, all aspects are visible at any given moment, which makes the measurement unproblematic. With quantum complementarity, measuring one part of a pair, such as the position of an electron, makes it impossible to measure the other, yet both are necessary for a more complete understanding of the object of study. As already indicated, Bohr challenged the classical notion of reality; the 'quantum of action' was indivisible, which suggested that not all aspects of a system could be viewed simultaneously (Frescura and Hiley, 1984).

Quantum physics not only shifts emphasis from things to relationships – or more accurately, to relationships in things – but also from being to action. According to the American quantum physicist Henry Stapp (2007: 20):

> To place action at the centre of reality upset the whole apple cart. It produced a seismic shift in our ideas about both the nature of reality and the nature of our relationships to the reality that envelops and sustains us. The aspects of nature represented by the theory are converted from elements of being to elements of doing. The effect of this change is profound. It replaces the world of material substances by a world populated by actions, and by potentialities for the occurrence of the various possible observed feedbacks from these actions.

Stapp's 'seismic shift' replaces material substance and being with action and doing.[6] Action is a phenomenon that is easily understood in macroscopic terms but is less straightforward at the microscopic level. Karen Barad (2007) develops an ontology of mattering as an active process by which agents, both microscopic and macroscopic, are shaped through intra-actions. In her account, action is explicitly relational. Action does not refer to the movement of a discrete agent but represents an engagement with the other and world through which both are transformed. *Inter*-action is a transactional exchange between 'beings' from which the separate parts emerge intact. By contrast, in the *intra*-action, each part is shaped by what Barad refers to as a 'cut'. The particular cut defines the difference between parts and their relationship within a whole. The cut is not singular but ongoing as part of a process of becoming. The parts are mutually constituted, which as Wendt (2015: 260) argues, is only tenable within a quantum framework. Mattering occurs at all levels of life and is a process by which agential cuts are enacted and boundaries of difference are drawn.

The parallel between complementarity and Daoist yinyang suggests a somewhat different angle on the problem of action. Like Barad, Daoism grounds the human in nature and the universe and emphasizes the ontological parity of all being. But like Bohr, the Daoist yinyang highlights the difficulty of observing polar opposites in motion at the same time. Yinyang thus brings together a mind/action relationship with a dynamic relationship between the seen and unseen dimensions of a context, the navigation of which is crucial to how action unfolds. The relationship between what is non-present and unseen (wave, yin) and present and seen (particle, yang) is a co-creative relationship. As with Barad's concept of intra-action, the agents or actants both shape and are shaped by interwoven parts and wholes. Complementarity and yinyang suggest an ontological parity of parts and wholes and the mutual constitution of polar opposites. The whole is not prior to the parts, as would be the case for a monad, nor do the parts exist in a dualistic relationship of either/or. Daoist philosophy sees contradiction as a generative force in the world, from which its diversity emerges. Like the wave and the particle, yin and yang as polar opposites are interpenetrated, and raise a question about how the relationship between the seen and the unseen relates to action.

Action that arises from yinyang rests on paradox or contradiction. Daoism has a concept of actionless action or Wuwei in Chinese. To say that action is actionless is not to say that it should be equated with inaction or passivity, as might be assumed. A focus on inaction or passivity ignores the dynamic nature of contradiction. Action that is actionless is attuned to the larger

[6] The elementary quantum of action was actually discovered by Max Planck in his work on black-body radiation, but has many meanings within physics.

environment of which it is a part, and is able to navigate the potentials, both seen and unseen, with minimal effort and to maximum effect. Strategy, as will be discussed in the next section, involves the navigation of polar opposites within a context. Like Buddhism, Daoism further reveals the ethical implications of contradiction and paradox. For instance, Lao Tzu, the purported author of the *Daodejing*, states in Chapter 2 that we only know beauty through the ugly, and we only know good because evil exists. We don't generally see both at the same time, just as we don't see the wave within the particle while focused on the latter, but they are mutually implicated and always in a changing relationship.

Many Western traditions of ethics begin with abstract reasoning based on first principles. In practice, good and bad are often projected as essences onto others, that is, others are by nature good or evil. From the perspective of yinyang, good and bad are *potentials* that exist within the universe, world and human selves. The good flows from maximizing potentials that empower, whereas evil is that which damages, interferes or harms the flow. For instance, industrialization has accelerated the development of human artefacts that are not only harmful but destructive of nature, which has created an imbalance that is played out in extreme weather events with climate change. Human technologies that instead work with nature would empower positive potentials. From a Daoist perspective, the issue is balance and harmony, which nature, left to its own, will restore. The key question is the extent to which human artefacts are in harmony with nature and co-create sustainable forms of life or are destructive of them.

A further implication regards the relational ethics of contradiction. Barad presents a concept of intra-action with others. The threading of parts through each other suggests an obligation to, or respect for, the other. Each cut into the relationship brings with it an ethical responsibility. Daoism adds a particular quality of relationship both between humans and nature and between humans based on deference. As is discussed in more depth in Snapshots 3 and 4, deference means acknowledgement and yielding to others within a system of which one is part.

Strategy

The idea that the speed of travel cannot exceed the speed of light is an important principle of Newtonian physics which has implications for the meaning of causality. For something to cause and have an effect on something else, the two must exist in a local relationship rather than far away. Entanglement suggests the potential for phenomenon to act on one another even when separated by great distance. Einstein referred to entanglement as 'spooky action at a distance' (Einstein et al, 1935), which suggests that even when separated, the movements of one phenomenon impact on another phenomenon with which it is entangled, such that when X moves up, Y moves down. Relationships in action are not purely local but may be non-local in space and time.

Entanglement is at the heart of interpretations of quantum physics (Barad, 2007: 253) and can take different forms. The observer is entangled with the apparatus. For Bohr, complementarity is about the entanglement of wave and particle. According to Wendt (2015: 257), 'When particles become entangled they acquire new properties, namely relational properties to the whole', such that parts and whole are 'co-emergent'. Barad (2007: 384) argues that ethics is 'about taking account of the entangled materialisation of which we are part.'

The Buddhist and Hindu concepts of karma, which means action, reveal several aspects of entanglement across time and space. The snapshots in Section III explore entanglement in practice. Snapshot 5 brings the parallel to the analysis of a structure of colonial power and the potential for the strategist, as a participant in creating the world, to transform the contours of a relational entanglement of power. Snapshot 6 moves to more systemic entanglements extending over hundreds of years and the reproduction of habits and relations of 'thing-ness' as expressed in the concept of a chattel slave or human as property. The concept of karma suggests that actions are not localized but resonate across time and space within a feedback loop that is potentially recurring. Karma is as such an activation that resonates non-locally. Its temporal and relational dimensions suggest that every movement contains potentials that either reproduce or transform systemic relations over time.

The question regards the implications of this entanglement, not least across time, for strategy. To strategize is a very human activity and one that easily connects to the subject matter of international relations. Strategy would not, however, seem to apply to subatomic particles. Yet Barad's (2007: 269–84) claim that even the brainless and eyeless brittle star 'navigates' its environment points to something like strategy that goes 'all way the down'. Mind and the qualities of intelligence, or seeing and knowing in this example, are not equivalent to the possession of a brain or eyes, respectively. Navigation, whether non-human or human, involves the ongoing *negotiation of difference* or contradiction, whether between inside and outside, self and other or human and nature. The Hindu *Mahabharata*, as suggested by Datta-Ray (2015), is ultimately concerned with the potential to navigate the world without doing harm. In this case, it is the tension between universal truth and human action in the world that is the subject of negotiation.

Navigation and strategy are not the same, but they may be entangled. Strategy, on its own, like agency, is heavily laden with connotations of autonomous actors and linear reasoning to achieve objectives. Strategy contains a further connotation of acting from self-interest. In her discussion of ethics, Laura Zanotti (2019: 122) argues that a Newtonian imaginary that emphasizes the separateness of actors and abstract reasoning is not sufficiently attentive to context and thus not only likely to get things wrong but to end up diverting responsibility. A navigational strategy, by contrast, requires the continuous negotiation of uncertainty that arises from the creative coexistence of polar opposites within a context that is both relational and emergent. Strategy is the activation of intention that exists not in a single moment but which percolates out to relationships that are spatial and temporal. Both the relational and temporal nature of entanglement provides a practical and ethical incentive to minimize harm. In harming the relational whole of which one is part, or other parts that are intertwined, the self is also harmed; likewise, care for the whole, or other parts within it, constitutes care of the self. Entanglement makes other-interest rational and pure self-interest dangerous to both self and other. Self-interest and other-interest are not mutually exclusive but entangled (Dalai Lama, 2000: 48), which has implications for how we conceive of what is possible, desirable and ethical for individual, social or political life. The navigation of uncertainty and relational difference is central to the thought of Sun Tzu and Gandhi, two of the world's greatest strategists and navigators. As the snapshots reveal, mind, action and strategy run through each other as conscious elements of activation, and intersect with qualities of self, structure and time.

Quantum physics tells us something about the nature of mind, action and navigation at the subatomic level. Exploring parallel conceptions in Buddhism, Daoism and Hinduism places human mind, action and strategy within a seamless universe, running from the microscopic through the macroscopic. Both involve a shift from being to becoming within a 'reality' that is lacking in material foundations as understood in classical physics. Table 1 illustrates Newtonian assumptions in contrast to the parallels between quantum physics and the Asian philosophies, as these relate to mind, action and strategy.

Contrasts and parallels

The parallel provides a tool for making a particular 'cut'. In a quantum world, action does not stand alone as a discrete thing that interacts with other things, nor is it the single movement of a being in local space and time – action is the activation or enactment of mind and movement in its unfolding, shaped by both intention and context. In unpacking a broader notion of action, I gradually replace a concept of agency with 'assemblage', a term made popular by Deleuze and Guattari ([1987] 2016).[7] The expandable

[7] While this is discussed further in Snapshot 4, my primary intention is to bring the concept to Eastern thought rather than develop the use of Deleuze and Guattari. (On more directly Deleuzian uses of the assemblage see, for example, Salter, 2013; Abrahamson, 2016; Bueger, 2018.)

Table 1: Contrasts and parallels

Contrast		Parallel: impermanence
Newtonian	**Quantum**	**Buddhism**
Independent reality	Real and non-real	Real and non-real (self)
Material things	Real: relations	Real: relations
Essence	Non-real: illusion of essence	Non-real: illusion of essence
Mindless matter	Mind in matter	Mind in matter
Observer separate	Entangled in language	Entangled in language
Is vs ought	Moral responsibility of the scientist	Turn from ego to entangled self
Contrast		**Parallel: complementarity/yinyang**
Newtonian	**Quantum**	**Daoism**
Being	Action	Action
Binary: non-being	Particle-wave duality	Yinyang
Presence	Non-presence/ presence	Non-presence/presence
Interaction of parts	Entangled intra-actions	Entangled intra-actions
Causes	Potentialities	Potentialities
Good and evil as essence	Moral ambiguity of quantum technology	Good and evil as potentials
Contrast		**Parallel: navigational strategies**
Newtonian	**Quantum**	**Asian strategies**
Strategy	Navigation	Strategies of navigation
Linear	Negotiating difference	Negotiating difference
Abstract	Context dependent	Context dependent
Agents vs structures	Participant in creation	Participant in creation
Local	Non-local entanglements	Non-local entanglements
Reasoning from first principle	Ethical responsibility for material entanglements	Minimize harm, transform relationships

nature of the concept makes it possible to observe activations of consciousness as they relate to articulations of the self who acts and the space and time contexts of intra-action. The focus on mind, action and strategy represents a repositioning of aspects of the social science debate, including Wendt's (2015) emphasis on consciousness and mind, the importance of mattering and navigation in Barad's (2007) account, and Zanotti's (2019) discussion of the relationship between context, agency and ethics.

Entangled in language

Bohr's epistemological revolution became a focal point of criticism, not least by Einstein (Einstein et al, 1935), who argued that the theory is incomplete, although experiments since suggest that Bohr won the debate with Einstein (AMNH, 2021). Both Barad and Wendt argued that Bohr's focus on epistemology was incomplete and sought to explore the ontological significance of quantum theory. Badredine Arfi (2018) has argued that Barad and Wendt's emphasis on ontology doesn't address the problem of quantum 'background independence' or the absence of metaphysical primitives. As ontology is about fundamental units of being, that is, metaphysical primitives, the tension is palpable.

In the classical world, 'reality' is composed of matter. It is made up of material objects and has a quality of 'thing'-ness. Once any notion of solid matter or essence dissolves, to be replaced with the indeterminism of particle and wave, there is no intrinsic being in the world and thus no fixed backdrop of reality. As already suggested, all is change and impermanence. Physicists have a mathematical language and procedures for conceptualizing background independence. One debate in the social sciences revolves around the use of mathematical insights as a conceptual framework for theorizing the social world (Arfi, 2018: 103). Arfi is critical of Wendt for failing to engage with the mathematical language of quantum physics. By contrast, Bentley Allan (2018b) argues that, even while he avoids the mathematics, Wendt's argument rests on a problematic claim that mathematical descriptions map onto human and social reality in a realistic way. As Allan (2018b: 97) states:

> The danger in accepting QDT [quantum decisionmaking theory], quantum linguistics, and quantum game theory as the core of post-classical social theory is that we will always be working from mathematical symbol sets to human action, instead of the other way around … We can adopt a contextual view of human agency without having to constantly work our account of the social into the terms and mathematics of physics.

Furthermore, he argues, Wendt remains within a scientific realist framework that is itself bound up in a classical episteme.

The debate reveals three important points. First, Bohr's parallel highlights the potential significance of quantum theory for other domains of experience. His formulation provides a natural stepping stone for asking about the significance of quantum physics for human observation and action. Second, Bohr, who inspired an epistemological revolution, refers to the usefulness of Buddhism and Daoism for exploring epistemological problems. In the realm of social analysis, an exploration of these epistemological problems, I would argue, requires neither mathematical formulae nor the translation of mathematics into everyday language. We do not need to do so because the relationship between background independence and the materialization of form was articulated millennia ago by Asian philosophers who reflected directly on the implications of a conception of the universe like that rediscovered recently by quantum physicists for human mind, action and strategy. Third, the parallel between quantum theory and the Asian philosophies makes it possible to situate human mind, action and strategy within a coherent model that builds on the paradox of real and non-real, rather than the binary of either analogy or actuality. If the world is holistic rather than divided into microscopic and macroscopic, the seeming matter and mechanism of the human world may be more a reflection of the apparatus through which it has been observed over the last few hundred years than an essential quality. In this respect, as Wendt (2015: 287) suggests, the perception that we are getting closer to truth is more a reflection of the role of repeated measurements in stabilizing a particular reality rather than in capturing an independently existing one.

As Barad and Wendt both contend, quantum physics changes the relationship between the physical and the social sciences. Buddhism, Daoism and Hinduism provide a point of departure for a transformative redrawing of boundaries between the two. The Asian philosophies also provide ethical insight that draws directly on human experience and its relevance to social life. The question is whether quantum social science should start by leaning into a field of science within which the social sciences have historically been the poor cousin, or by leaning into alternative frameworks of thinking that have historically engaged with problems of human mind, action and strategy, but which have been ignored within the context of modernity because they are not of the West. The answer need not be either/or but can instead be either/and.

The ancient Eastern philosophies provide a backdrop for conceptualizing mind, action and strategy in a manner that gives humans a home in the universe. Social scientists do not need to look to the mathematical language of quantum physics to understand this human home. If instead we start with a recognition that mathematics *is* a language, albeit one more formal than everyday language, then it is the *language* of the observer that brings a particular context or world into being, whether at the level of sub-atomic particles or human systems.[8] Rather than a hierarchy between mathematical and everyday language, the central problem, as Bohr suggested, is one of epistemology and the relationship between language and world. It is further about the materialization of form in the absence of an ontological background. Epistemology cannot ultimately be separated from ontology, although it may be difficult to observe both at the same time (Fierke, 2019).

As Colin Wight (2018: 155) notes, 'When we try and transpose the mathematical clarity and coherence into a context of meaning, we impose on the ineffable a set of paradoxical outcomes.'[9] In any case, the transposition of human contexts of meaning into a mathematical language is unnecessary, and indeed would only introduce distortion. Rather than bringing the language used to describe sub-atomic particles to the analysis of macroscopic phenomena, we might instead start with philosophies and practices that have arisen from other times and places, which, Heisenberg notes, illuminate the dynamics of human life in a quantum world.

The physicist relies on the language of mathematics to make sense of the activity of subatomic particles. Physics is thus no less confronted by questions of language than the social sciences. What is the relationship between language and world? Does language represent a physical reality which exists independently of it, or is language bound up in and constitutive of realities? What is the relationship between the language of the scientist or analyst and the object of study? A conversation between physicists and social scientists might explore these common questions of language. A shift from an understanding of language as representing an objective world which stands apart to one that sees language as bound up in the process by which worlds are known and formed is fundamental to the shift from Newtonian materialism to quantum consciousness and mattering. In turning from the Newtonian to the quantum, the binary of either epistemology or ontology gives way to the mutual implication and inseparability of epistemology and ontology.

As Arfi (2018) claims, Wendt and Barad are unable to address background independence because of their focus on ontology. Contrary to Arfi, we

[8] For a deeper interrogation of the relationship between Wittgenstein's philosophy of language and mathematics and physics, see Bitbol (2018).

[9] As Smetham (2010: 26) notes, despite all of the technology which rests on knowledge of the functioning of the quantum realm, physicists remain in 'the dark as to what *really* is going on beyond the probabilistic predictive mathematics; they certainly do not know what "exists" at the quantum level'. As Richard Feynman said, it is 'safe to say that nobody understands quantum mechanism' (Dalai Lama, 2005: 64).

understand mathematics as a language which is more formal than everyday language, but a language nonetheless. The ancient Asian philosophies provide an understanding of the relationship between the formlessness of background independence and form, of which language is one expression which connects the micro- and macroscopic in a seamless whole. This radical departure from Newtonian thinking about the physical world moves us away from a focus on contained intrinsic subjects, objects or measures of time.

Science and mysticism

In the use of language, wave functions collapse into form. Science and mysticism, as words that are deeply sedimented in our classificatory system, evoke an immediate recognition and convey a set of assumptions that are often embraced or discarded without looking further. The question is the extent to which, as Wittgenstein (1922: 5.6) – or Bohr – might say, we are trapped – or suspended – in our language and unable to see beyond it.[10] For instance, the category of the human, and what is defined as not-human, can itself frame the context within which particular problems are defined. Ip Po-Keung (1983) points to a definition of nature as a non-human environment. While such a formulation might be taken for granted, an entire philosophical system and assumptions about how to engage with the world emerge from it. The conceptual difficulty extends further, to, for example, what constitutes metaphysics and to what extent the term relies on a dualism (Ames, 1986: 324) that is contrary to the nondualism of Buddhism and Daoism. If metaphysics assumes first principles, can the term appropriately be applied to a universe characterized by impermanence and background independence?

The conceptual difficulty that arises from the contrast between science and religion is a further case in point. The concept of religion relies on some notion of the transcendent, usually associated with a deity, as well as a metaphysics, which constitute a dualism between world and a universe that stands outside it. On the one hand, the dividing line between world and that which transcends it is indebted to modernity; on the other hand, most religions connect us to something beyond the self, as well as the potential for compassion towards others, even while recognizing that compassion is not ubiquitous. It is useful to think about where religion, metaphysics, cosmology or ontology begin and end, and why we want to use any one of these concepts. To what extent does any particular use belong to a contextually specific language game?

10 'The limits of my language mean the limits of my world.'

The concept of religion is a contested concept like any other, and not least as regards its application to, for instance, Buddhism or Daoism. Having said this, Clifford Geertz (1968: 97) made a point about the 'religious perspective' that is apt. He claimed that at its heart it 'is the conviction that the values one holds are grounded in the inherent structure of reality, that between the way one ought to live and the way things really are is an unbreakable inner connection.' One's perspective on the world's construction is inseparable from conduct within it, that is, human conduct contributes to a particular construction of the world, which then shapes further conduct within it. From this notion of religious perspective, the shared values of ascetic Protestantism and capitalism, identified long ago by Weber ([1958] 1976), might be explored in relation to the first principles of Newtonian science, including the division between religion and science assumed within this constellation. As Thomas Kuhn ([1962] 1970) has shown, science is no more monolithic than religion. The ability to recognize and embrace other forms of science has implications for how the relationship between science and other values is conceived.

The particle-wave relationship in quantum physics means that the fundamental constituents of either the world or the larger universe are no longer 'things', defined by what they are not, in terms of either/or. Instead, particles and waves exist in a relationship that is complementary, such that any one particular is particular but can also be understood and explained in its dynamic relationship to a polar other. A relationship that is dynamic, rather than static, is impermanent because it is in a constant state of change. There can be no predetermined pattern or set principle, such as would be found in an underlying mechanism, which is prior to and forms the background against which any particular or local figuration can be understood. Instead the pattern emerges from indeterminism and change, as concrete particulars converge into intricate patterns of difference. The result is more akin to the harmony (or disharmony) of an orchestra or the preparation of a stew, where the particularity of the participants or the ingredients, respectively, emerges from a pattern of difference in harmony (Ames, 1986: 322).

The big question, which is crucial to any claim that quantum physics is relevant for the social sciences, regards the move from the activity of particles and waves to that of humans or other sentient life and why we might want to avoid the language of metaphysics, religion, mysticism or even cosmology in approaching this different angle on the world. If we are suspended in language, as Bohr said, the categories that we use habitually must be temporarily suspended in order to engage with a conception of reality that is otherwise too quickly dismissed precisely because the categories are nested in a binary and hierarchical relationship to science. A fundamental distortion and tension arises from maintaining the dualism. Wendt (2015) expresses this tension in his claim that constructivists, no less than other social scientists, cling to Newtonian first principles even while trying to move beyond them. They cannot, as a result, provide a physical account of mutual constitution. Arfi (2018) suggests that Wendt makes a similar error in using the language of quantum metaphysics and ontology.

Newtonian first principles start with 'things' or essences, characterized by their independence, materiality, locality, determinism and mechanism, such that interaction with an other is purely extrinsic and leaves each component part intact (Allan, 2018a; Kurki, 2020: 3). Quantum theory relies on a more holographic conception of relationality, characterized by mutuality, indeterminism, complementarity or correlativity, which means that any notion of self exists in a polar – rather than a dualistic – relationship with the other. The self, at one and the same time, exists in its particularity but is continuously shaped and transformed by encounters with the particularity of others, as manifested in an ever-changing context.

By way of destabilizing the dichotomy between Newtonian matter and quantum consciousness, and to rethink the relationship between them, a snapshot of Sir Isaac Newton himself is revealing. Newtonian physics is associated with pure materialism and objectivity, the idea that the universe is a mechanism, and in Wendt's account, the absence of attention to consciousness, which is fundamental to any sense of being at home in the universe. Social scientists, he argues, have an ontological grounding in Newtonian assumptions. What happens in the shift to the relationship between consciousness and matter? The rest of this book is an attempt to explore what that shift might mean in practice. The brief sketch that follows suggests that what Newtonian physics has become rests on an incomplete version of Isaac Newton's thought. It includes only one side of the equation. The side of him that connected to a notion of consciousness or intelligence in the universe, or of God, is hidden but nonetheless there, infusing his articulation of Enlightenment principles.

Newton, the man whose name has come to represent a conception of the world that is pure matter and mechanism, was more complex. He was not only a scientist but an alchemist, and his writings, including his unpublished works, contain more documents on alchemy and theology than science.[11] Newton had a huge influence on Enlightenment thought, not least that of Galileo, Descartes and Adam Smith, but he himself had two sides. The visible side was the scientist. The hidden side was the alchemist.[12] The silence

[11] One million, four hundred thousand words regarding theology, 550,000 on alchemy, 150,000 on monetary affairs and a million on science (Verlet, 1993).

[12] Alchemy, which was a medieval predecessor of chemistry, but which was closely associated with mysticism, was concerned with the transmutation of matter and attempted to transform base metals into gold or to find a cure for disease and extend life.

surrounding the latter was only broken in 1968 (see Manual, 1968). The two sides did not exist in an either/or relationship but were interpenetrated, not least in his work on the Enlightenment (Westfall, 1980, 1993).[13] Newton's unpublished manuscripts, discovered two centuries after his death and purchased at auction by John Maynard Keynes in 1936, revealed that Newton was the 'last magician' as well as the 'first physicist' (Bauer, 2007: 60).

The conceptual system that has arisen from Newtonian physics does not have a place for consciousness, unlike its author, who did. Consciousness is not the same as alchemy, but both express possibilities beyond the assumed causal closure of classical physics. Newtonian physics does explain a great deal; however, it is not in itself the complete story, and much of its baggage has been scientifically discredited or surpassed (Kurki, 2020: 3). Given that all is matter from a Newtonian perspective, it is not possible to see or account for quantum consciousness. However, once the observer repositions the apparatus, and is able to see consciousness in matter, the Newtonian world remains part of the equation. The fact that Newton felt the need to keep his mystical interests secret in itself suggests the presence of other cultural trends or power relationships that were contributing to the constitution of a purely material world, to which his science then contributed. When the hidden and complementary dimensions of Newton's thought come into view, the emphasis on materiality in his physics is transformed from universal 'truth' into to a localized configuration arising in time and space that shaped a particular historical context.

Milja Kurki's (2020: 7) cosmological analysis is an attempt to reveal how conceptual systems are aligned with certain understandings of the cosmos as well as specific understandings of God, science, physics and philosophy. The prevailing clear separation of the social sciences, including international relations, from the natural sciences arises from the legacy that Newton shaped. The further claim of the present book is that the ancient philosophies which are more akin to quantum physics may assist in repositioning the apparatus, which might also then, as Kurki (2020: 7) further notes, enhance the potential 'to think "social" science again and think again *with*, rather than apart from, the "natural" sciences.' Such a reorientation makes it possible to look at how a framework for 'knowing' has contributed to some of the seemingly intractable problems we face, which might, after repositioning the apparatus, have greater scope for resolution (Wendt, 2015: 32).

[13] The complementarity of his science/mysticism relationship is reflected in Newton's use of an anagram, 'jeova sanctus unus', as a pseudonym for Isaacus Newtonus. The anagram, which is the opposite construction of his name, made it possible for him to correspond anonymously with Rosicrucians and alchemists (Bauer, 2007: 62).

From a quantum perspective, the seeming fixity of the modern world arises not from underlying principles or laws of nature but rather from categories that have constituted and informed a series of practices within the world during a particular historical juncture. These categories have defined the world and the universe as a mechanism, divided up into discrete territories. The metaphysics have legitimated carving up the global space into things called states and subjugation within that space to a pre-assigned pattern of relatedness, which have emphasized conformity to laws of nature and rightness as a function of this conformity. Conformity was imposed through the process of subjugating other cultures. Whether looking at the Highland Clearances in Scotland, the displacement of Native Americans or Australian Aboriginal peoples or the displacement of Africans to enslavement in the Americas, the demarcation of boundaries and states went hand in hand with subjugation of the particular, from the speaking of local languages to the wearing of native dress – for example, the Tartan – to architecture – as Black Elk, a heyoka of the Oglala Lakota people, said, no more round houses, only square ones (Neihardt 2014) – to the epistemicide of local traditions, all in the name of conformity to the mechanism of modern life.

The point here regards the co-constitution, and thus the inseparability, of science and religion or culture within a particular historical configuration which has shaped the world as we presently know it and relies on a particular cosmological vision (Allan, 2018a; Kurki, 2020). From a quantum perspective, it is possible to see how the science/mysticism dualism was built on a perception of mutual exclusivity, not least of races, while acting as two entangled parts of a hologram that informed practices that fundamentally transformed the world. Practices change across time and space, and mutually implicated oppositions are often the underlying dynamic of this change. The point should be obvious on the back of Kuhn's ([1962] 1970) argument about paradigm change. But the relationship between the Newtonian and the quantum is less that of a paradigm shift than a paradox, more akin to Wittgenstein's (1958: 194) duck/rabbit picture. The relationship between opposites is not either/or but either/and, even while only one half of the paradox is visible at any one moment.

The exploration of mind, action and strategy in an uncertain world involves a repositioning of the apparatus to look at the world/universe from a different angle. The parallel provides a way to think about the social implications of this reorientation, not in the sense of going back to the future, or returning to the cultural origins of Daoism or Buddhism, but rather of questioning our own foundational assumptions and the foundational sciences to which they belong, much as Einstein or Whitehead did. Such a reorientation might also, as suggested by Roger Ames (1986: 320), position us outside of Western traditions of thought and practice in order to more clearly view the assumptions contained within them from more neutral ground, thereby giving us a new hermeneutical sensibility. Ames's articulation, more than thirty years ago, contains a distinction between 'our' position or approach and that which might lie beyond, an alternative. With the repositioning, West and East, or North or South, can no longer be – if they ever were – conceived as fixed places and contained cultures that 'own' specific sets of ideas.

Capitalism and Christianity have a presence in Asia just as manifestations of Daoism and Buddhism are evident in the West, if in different ways. Resonances of the assumption that 'West is best' remain, but a broader array of conceptual resources are needed to address fundamental questions about what it means to act in an uncertain world, a world that is being pulled apart by increasingly deep political divisions, inequality, climate change and pandemic. The mutually exclusive distinction between science and mysticism, or secular and religious, is not suitable for addressing the challenges that we face. As Einstein noted, we cannot solve our problems with the same thinking we used when we created them. Until this is accepted, we will continue to work at cross purposes with ourselves.

Science and culture

The post-colonial critique and the move toward a more global international relations have highlighted the Eurocentrism of international relations theory and the need to bring more voices to the table, even while this is complicated by the entanglement of the contemporary world with modern structures that have their roots in colonialism and imperialism. Thought systems from other parts of the world have not been sufficiently taken into account within international relations scholarship, although, as discussed earlier, this is changing. One of the difficulties, as discussed in the opening section, is the tendency to relegate any system of thought that differs from the assumptions of unitary science to the category of culture, where it can be dismissed. Another literature, arising from the social sciences more broadly, makes a different point that the philosophies that underpin Eastern practices, such as mindfulness and yoga in particular, or adaptions of Sun Tzu to corporate practice or military strategy, have been 'lost in translation' as they have travelled West. The cultural appropriation of mindfulness or yoga, for instance, has, it is argued, transformed these practices beyond recognition (Mitra and Greenberg, 2016; Hsu, 2016; Titmuss, 2016; Walsh, 2016).

Claims that Eastern practices have been victim to cultural appropriation by the West present an interesting contrast to critical arguments regarding travel in the other direction. Mills (2014), for instance, highlights the extent to which Western models of psychiatry are being globalized, subsequently colonizing global mental health and delegitimizing other cultural conceptions of healing. In the process, the extent to which Western medicine itself rests on cultural assumptions tends to be ignored (Watters, 2010). Similar arguments have been made about capitalism and the global economy. The adoption of techniques of Asian origin by Western institutions, from hospitals to schools, universities, corporations or the military, suggests that the 'travel' is multi-directional. But the travel is also across time, given that 'modern' Buddhism, which is one label for the Western variation, is itself a by-product of earlier colonial interventions. The 'scientific discovery' of Buddhism was made by Western scholars in the context of colonial expansion. These scholars adopted early Theravada Pali texts as the expression of the universal essence of Buddhism (McMahan and Braun, 2017: 6). It was this version that provided the basis for a 'world religion' that was born in Europe (Lopez, 1995: 2) but which erased much of the ancient cosmology of Buddhism. Modernizers from Asia then later transformed meditation into mindfulness, underpinned by modern psychology, thereby decontextualizing and separating it from its association with, not least, the Eightfold Path that forms the core of Buddhism (Walsh, 2016).

As Robert Purser (2019: 85) notes, not all cultural appropriation is negative. The cultural appropriation of Buddhism in China was different in kind than acts of cultural appropriation by the Western imperial powers in modernity. As it arrived in China, Tibet and Southeast Asia, along the ancient Silk Road, Buddhism more or less peacefully mixed and coexisted with other more indigenous traditions, such as Daoism, rather than writing over them, although in China these traditions were much later suppressed in the context of Mao's modernization and Cultural Revolution, or more recently in Tibet. Purser's point concerns the epistemic violence of the cultural translation of modernity, and the complex set of interacting power relations, networks of interests and interpretative decisions that are often hidden in public discourse. While Buddhism, Daoism or Hinduism coexist and overlap in different Asian cultures, as is evident from the snapshots that follow, the emphasis in Western thought on identifying the one true reality has gone hand in hand, in the context of imperialism, with the construction of clear hierarchies of difference and often the destruction of indigenous culture.

The big question at the heart of the debate about cultural appropriation is whether Buddhist mindfulness or yoga, with roots in Hinduism, among other practices, when introduced in the West have become, in a very Foucauldian sense, mechanisms of neoliberal discipline and control, or whether they hold out the potential for more institutional and contextual transformation. In 2013, the year before declaration of the 'mindfulness revolution' by *Time* magazine, the UK *Economist* claimed that 'Western capitalism seems to be doing rather more to change eastern religion than eastern religion is doing to change Western capitalism' (as cited in Titmuss, 2016: 191).

The debate, and arguably the longer history, revolve around a contest between two different conceptions of reality. Much like Capra's distinction between Eastern mysticism and Western science, and as revealed in the quote in the previous paragraph, the critical literature on cultural appropriation by the West tends to equate the Eastern traditions with religion. Yet, the meaning of religion is neither straightforward nor fixed. While there are religious variations of Daoism and Buddhism, and Hinduism is more explicitly religious, these are also philosophies or cosmologies that relate to systems of inquiry, and the latter, at least in the case of Daoism (King, 2015) and Buddhism, preceded the religious expression. The parallel to quantum physics suggests it may be more fruitful to place both Buddhism and Daoism within a broader notion of philosophy or cosmology, which shifts emphasis away from a deity to the nature of the universe and becoming within it. Kurki (2020: 8) argues that a relational cosmology abandons a 'God's eye view of sciences.' Sciences are no longer about capturing the truth but rather are an attempt by situated knowers to 're-approximate' the world. From this perspective, science becomes philosophy and philosophy becomes science, both of which are inextricably bound up with politics and how we think about society and ethics.

A further distortion arises from the conception of culture, in some of this literature, as a container which can be neatly aligned with one spatial context or category of people but not another. In this formulation, the main debate regards the appropriation of culture by those who have no claim to it, and for their own purposes, in a manner that is no less problematic than colonialism (Snyder, 2006: 101). The conceptualization relies on a contrast between a practice, as it emerged within a specific culture, and the assumptions and objectives to which it has been applied when translated into a new context. The problem is twofold. First, while Buddhism originated with the Buddha, in a particular time and place (that is, Northeastern India in the late sixth century BCE), and contains assumptions that are indebted to Hinduism, while also diverging from it, practices such as meditation have evolved and taken different forms across Buddhist cultures. Second, the epistemological conflict between Western individualism and Buddhist ideas of self concerns a difference in kind which arises from diametrically opposed assumptions.

As traditions move across borders, away from their context of origin, they do undoubtedly change, and often beyond recognition. But this has also given rise to a conversation about those roots, which is potentially transformative. The point is that to treat either culture or religion as containers is to fall into a modernist trap of ownership, intrinsic identity and illusion, which obscures the extent to which ideas and practices, like people, travel and are often transformed by the journey. The sheer number of Chinese or Indians or Africans or, for that matter, Europeans who, across several generations, have lived in America, and who themselves have multiple cultural experiences, already highlights the problem. People, like theory (Said, 1983), travel from a set of initial circumstances of birth to a new location where they are confronted with conditions specific to that place, and, one might add, their position within that place. They adapt in ways that are conducive to acceptance, but then are transformed through their accommodation to a new time and place. Alternatively, on what basis would it be possible to make a judgement regarding the essence or truth of various schools within Buddhism or Daoism that would *not* be in conflict with the notions of self and truth upon which these traditions rest? Indeed, Buddhism, Daoism and Hinduism recognize the limits of our conceptual world, both celebrating multi-perspectivity and acknowledging that the boundaries drawn in language can be the first step toward conflict and war. As Wang Chen notes in the classic *Dao of Peace* (Sawyer, 1999: 18–19), the cause of suffering can be found in the human tendency to conceptualize. He states that 'as soon as things have names and people emotions, love and hate arise and attack each other, warfare flourishes.'

Conclusion

Bohr or Heisenberg's parallel challenges the exclusive identification of Buddhism, Daoism or Hinduism with either religion or a specific culture, suggesting instead that they, like quantum physics, convey an understanding of how the universe and 'reality' – albeit as both real and non-real – more generally work. The Eastern philosophies provide an explicit place for humans and their practices within the universe. To explore this parallel is to deny neither the origin of these thought systems in a particular place and time nor their contribution to the development of particular cultures, but rather is to investigate the implications for mind, action and strategy in an uncertain world. Positivist social science and post-structural approaches have for the last few decades often been pitted against each other in an either/or, science-or-'reflectivism' struggle. Exploring the implications of the quantum turn through Buddhism, Daoism and Hinduism offers more space for understanding the deep constitution of life and intelligence in an emergent physical world.

In conclusion I want to highlight some issues regarding the incompleteness of the analysis. In preparing the manuscript, I have been confronted with a similar question from two different angles. From one angle, I have been asked why I haven't said more about other non-Western traditions, such as Islam or indigenous thought, where parallels to quantum physics might also be found. From another angle, I have not sufficiently explored strands of Western social theory or philosophy that resonate with quantum theory, or, in using examples of Newtonian thinking and practice as a foil, I have not accounted for the diversity of Western thought. While I would challenge any suggestion that we should not look beyond the West if ideas have already been explored by extant Western theories or philosophies, which is a comment that is sometimes heard (Wang, 2013: 35), I also do not mean to imply that possessive individualism and rational action are the only game in town in 'the West'.

No book can deal with everything. Trying to say something meaningful about the parallel between quantum physics and the Eastern philosophies as it relates to practice in the context of a global pandemic is already a huge task. To introduce further comparisons with theories or philosophies that share a family resemblance would be too big a remit for one book. The obvious reason for the focus on Buddhism, Daoism and Hinduism is the parallel drawn by quantum physicists to these particular traditions. The parallel has been articulated in different ways, but its frequency suggests potential insight into the significance of quantum physics for life and a home in the universe. By placing this concern at the heart of debates about, on the one hand, the relevance of quantum theory for the social sciences and, on the other hand, global international relations, the hope is both to contribute to and encourage a broader and more global conversation.

Here again I refer to the quote from the *Rig Veda* with which the book began: 'Truth is one, we call it by many names.' Once one moves from a conception of science that seeks truth in an independently existing reality to an exploration of the formlessness/form relationship and continuous processes of becoming, we are confronted with the inevitable incompleteness of any account of the world. Science is no more monolithic than social theory or human practice. Even within quantum physics, there are various points of theoretical disagreement. David Bohm (Horgan, 2018) made a similar point about the impossibility of a theory of everything and the incompleteness of knowledge, although some physicists would disagree. We are always creating and recreating the world, including our knowledge about it, and are thus in the realm of family resemblances rather than absolute truths. Likewise, within Buddhism, Daoism or Hinduism there are multiple traditions and practices, many of them formed within distinct cultures. The contrast between Newtonian and quantum assumptions highlights the difference between a world where only matter matters and one in which consciousness and matter are bound up with each other. The parallel between quantum physics and the Asian philosophies provides an apparatus for exploring the implications of a quantum turn for how we live as humans. It further highlights both the importance of and potential for greater diversity and thus the need for conversation about how to address the social inequalities, among others, raised by the pandemic.

From a political perspective, some readers may confuse an approach informed by philosophies with their origins in Asia with taking sides in the political debate over the rise of China. This would be a mistake. Both Daoism and Buddhism have a long history in China and are a part of the culture; however, and not least since the Cultural Revolution, these mix with other knowledge systems from communism to capitalism or Christianity. The latter has increasingly established a place in mainland China, while more ancient traditions, and Confucianism in particular, have re-established a place in political discourse. As Asian scholars have highlighted, China is a hybrid phenomenon (Shih and Jin, 2013; Wang, 2013, 2017), influenced by the forces of modernity and state sovereignty as well as its own more indigenous traditions.

Exploring the meaning of mind, action and strategy in ancient texts is not equivalent to a political position in the present. If anything, as explored in the conclusions of the book, the exploration might have something to say about how the US and China might rethink their relationship as well as the habits of 'statehood' in a context where global cooperation is needed to navigate the way forward. The book arose from the context of a pandemic, which exposed layer upon layer of destructive practice and habit against the backdrop of a tremendous loss of life. It asks how we might begin to acknowledge those habits and move towards new potentials. As centres of global power for the last few centuries, the United States and United Kingdom deserve particular attention. Those who have benefited from their positioning within this global order, including myself, have a particular ethical responsibility. The objective is less one of seeking truth in a theory of everything than gaining some insight into the constitutive features of mind, action and strategy in an uncertain quantum world. The snapshots demonstrate that human and other life is entangled with the larger universe, rather than standing outside and separate from it. These are thus snapshots from home.

SECTION I

Impermanence

SNAPSHOT 1

Self/No-Self

> There is a possibility that one will become tormented when something permanent within oneself is not found.
>
> *Majjhima Nikāya* 1.22

Initially, the coronavirus seemed to be an equalizer. It didn't recognize borders. It didn't distinguish between rich and poor. Everyone was equally vulnerable to the disease and could suffer from it or die. The unexpected and unusual susceptibility to suffering on the part of the wealthy created a new, if temporary, awareness of those who suffer daily, whether through illness, disability, poverty or homelessness. The sensitivity was elevated by the emergence of a medical frontier to tackle the virus. The frontline was occupied by many workers on the lower rungs of the economic ladder, now tasked with saving lives while potentially sacrificing their own. As time progressed, and demographics related to the disease came to light, the initial claim that the virus was an equalizer became suspect. Among the disparities, the old were more likely to die than the young, although the latter became more vulnerable as the virus mutated. But the most jarring reality related to the disproportionate number of fatalities within the US and UK who were of African, Latinx, Native American or Asian heritage. As the virus spread globally, poor countries were most vulnerable to the economic consequences of the pandemic, although many, not least in Asia and Africa, were, at least in the initial stages, more effective than the US or UK in managing the pandemic itself.

None of the societal or global divisions exposed by the virus were new. What was different, in the context of the pandemic, was the greater difficulty of pretending that the inequality of suffering wasn't there or of blaming the sufferers for their experience. The unusually large number of deaths within a short span of time highlighted the impermanence of life. As Hobbes ([1651] 1958: 185) famously said, life is 'solitary, poor, nasty, brutish and short.' Hobbes' *Leviathan* ([1651] 1958) provides an important foundation of Western realist and liberal thought, and contains assumptions that stand in stark contrast to those of Buddhism. Hobbes and the Buddha would be considered very strange bedfellows. But they did share one thing in common: both began with a concern about the impermanence of life while pointing to very different ways to address the impermanence. Hobbes wanted to escape the impermanence, and the suffering it caused, although he retained scepticism about the extent to which this escape was possible. The Buddha turned toward the suffering and saw the ability to embrace it as key to liberation from it. If Hobbes has been associated with realism, and a notion of reality 'as it is', the Buddha reveals the illusion surrounding this claim. In a world of impermanence, we are not 'as we are', but are rather both real and non-real.

Our positioning towards worldly suffering forms the heart of the Buddha's story. Siddhartha Guatama was the son of a king who ruled lands in the area that today forms the border between India and Nepal. As a young child, the Buddha was prevented by his father from venturing beyond the grounds of the castle. The father hoped to shield his son from the pervasive suffering in the world. Herman Hesse's novel *Siddhartha* (1951) reconstructs the Buddha's journey from the safe walls of the castle out into the world in search of meaning. The novel records three stages in this journey. In the first, the young Siddhartha became an aesthetic and sought transcendence through self-denial and poverty. He then sought meaning in earthly pleasures and the materiality of the world. Having failed to find what he was looking for, and in despair, Siddhartha sat next to the water. Suddenly he heard and saw in its movement the unity of all life: 'All the voices, all the goals, all the yearnings, all the sorrows, all the pleasure, all the good and evil, all of them together was the world. All of them together was the stream of events, the music of life' (Hesse, 1951: 136). By this time, Siddhartha had become a ferry man, helping people to navigate and cross over the river. He was both embedded in nature and the world and connected to something more than the self. Buddhism, in this account, is a middle way between the materiality and transcendence of the world. It is a turn toward acknowledgement of the impermanence of life and its interconnectivity.

The Four Noble Truths, which form the core of Buddhism, highlight the suffering of sentient beings. Suffering gives rise to craving, desire and attachment. Letting go and detaching from 'self grasping' puts one on a path that leads eventually to the cessation of suffering and compassion. Quantum physics says nothing about suffering per se. How can there possibly be a parallel between the two? The parallel is metaphysical, although, given its emphasis on first principles, metaphysics as a concept is not entirely suitable to describe a universe of continuous change. The parallel regards the impermanence of life and its implications for how the human self is understood. The theoretical physicist Carlo Rovelli (2018) draws on Buddhism to explain the emotional dimensions of this impermanence as it extends to the self and suffering. Thermodynamic decay, and thus impermanence, is a part of all life. An emotional burden arises from the difficulty of accepting our vulnerability and mortality and our limited time as participants in the world. Time is suffering. Birth, decline, illness and death are suffering. Being joined to that which we hate or separated from that which we love is suffering. '[The] failure to obtain what we desire is suffering. It is suffering', Rovelli (2018: 164–5) contends, 'because we lose what we have and are attached to. Because everything that begins must end.' The origin of our suffering is the illusion of permanence. The cause of suffering isn't in the past or the future but in the now, in our memory and expectation. From a Buddhist perspective, it is impermanence, and the nature of material attachments, that creates the conditions for suffering, as well the potential for liberation from it. While there is a pull to turn away from the suffering, Buddhism is about acknowledging it. The problem of impermanence is emotional. The ability to experience sadness and grief and acknowledge suffering and death, or, alternatively, to turn away from them, is fundamental to the constitution of the self who acts and the ethical implications of action. If matter is not of essence itself but is continuously decomposing and recomposing, this changes the nature of the reality with which we engage.

Impermanence provides a point of departure for thinking about the nature of the self. All humans possess an experience of 'I', yet this experience is often confused with an essence, whether in regard to our materiality or human nature. It is the sense of 'I' and what is 'mine' that gives rise to a desire for security and well-being, which we can never as living organisms achieve, precisely because of our impermanence. For this reason, the pursuit of an unchanging core self becomes a source of suffering and dissatisfaction. This first snapshot zooms in on the significance of these claims in order to explore the paradoxical concept of self/non-self within Buddhism. The concept of 'self' in Western thought usually presumes the existence of a 'soul' or 'ego' that is intrinsic to the identity of the 'I' and the source of its continuity. The Buddhist concept of self, by contrast, accepts that we all carry with us a sense of 'I' but that this is lacking in essence.

Buddhism subsumes a tension between experience and essence. Buddhism has many different expressions, shaped by numerous distinct contexts and traditions. Buddhist philosophy and practice have travelled across time and have been shaped by different cultural forms. Having said this, the Buddha's story and the Four Noble Truths are common to any tradition that carries the name Buddhism. The concern here is less with any one cultural expression of Buddhist practice than with gaining some insight into the philosophy that underlies these formative observations, and with sharpening these insights through an exploration of a practice that has taken hold within Western culture that shares a family resemblance with Buddhist meditation. The central tenets of Buddhism stand in stark contrast to the Western emphasis on materiality, the primacy of the individual and the satisfaction of individual desire, otherwise known as utility maximization.

The snapshot opens with the broad contours of a debate surrounding the growing popularity of mindfulness practice in the West and particularly the United States. My objective is less to weigh in on the mindfulness debate itself than to use it as a foil for making an argument about the relationship between meditation and language. The contrast between an emphasis on materiality, individuality and profit, so central to Western culture, and the assumptions of Buddhism become clearer though the juxtaposition. The second part starts with Hobbes' theorization of the unceasing motion that arises from the state of nature, driven by desire and aversion, and similar ideas in Buddhism. Both start with the problem of impermanence, but each gives rise to a different conclusion. Hobbes finds some relief from the constant motion in the creation of the social contract, and thereby elevates the problem to the relationship between states. The 'levels of analysis problem', as understood within the study of international relations, begins with this construction and forces us to choose between a focus on individuals, states or system, either/or.[1] Part three begins with the Buddhist concepts of dependent origination and emptiness, and the illusory 'I', expressed as a narrative self, which flows more seamlessly between individual and collective. The performance of 'I' and 'mine', at whatever level, arises in language. Part four then asks whether the cessation of the proliferation of concepts, one objective of meditation, has relevance beyond the individual and what this might look like.

Consciousness finds expression in form, and in language in particular. The mindfulness revolution has given rise to a proliferation of words, as demonstrated in the advertisement of some one hundred thousand books through Amazon (Purser, 2019: 13), which is arguably in conflict with one objective of meditation within Buddhism, that is, to bring an end to the proliferation of words and to calm the mind. From this perspective, the debate itself reveals a problem of *language*. Mindfulness is about consciousness. Consciousness expresses patterns of self. Behind the pattern there is no essence, only emptiness. As stated by the physicist Max Tegmark (2015), 'Consciousness is the way that information feels when it is being processed … It's not the particles but the patterns that matter.' Consciousness and the patterns by which it is expressed are most important, rather than

[1] The International Studies literature often refers to the 'levels-of-analysis problem,' which at its most basic refers to whether analysis focuses at the level of the individual, the state or the international system. See the seminal piece by Singer (1961).

impermanent matter. The snapshot closes with a further exploration of the workings of the collective 'I' in the context of the UK decision to depart from the European Union. The example provides unusual insight into the process of moving from competing narratives of 'I' to an illusory acceptance of one narrative.

Impermanence and self

In 2014 *Time* magazine coined the term 'mindfulness revolution' to refer to the sweeping popularity of mindfulness in the US and the West more generally. Many have argued that adaptation of a practice with origins in Asian Buddhism to Western society is a cultural appropriation (for example, Rosenbaum and Magid, 2016; Titmuss, 2016). Its translation into a very different context generates all kinds of tensions if not contradictions. A central concern of the critics is that Buddhist methods have been 'lost in translation' (Compson, 2014: 274) as they are applied in a totally different cultural context. The cultural appropriation of mindfulness has, it is argued, transformed the practice almost beyond recognition. Others respond that the practice is not meant to be Buddhist and thus is not an appropriation. My objective is to examine the broad contours of the debate itself, and what it says about the nature of Buddhist mind or 'self'.

The mindfulness revolution has swept the US and Britain, with the practice taking hold not only in corporations, such as Google and Amazon, but also in schools, hospitals and the military in both countries. Even the British parliament introduced mindfulness training for MPs (TMI, 2019). The revolution is scientific, seeking to understand the influence of mindfulness on the brain, as well as institutional and practical in providing a method which, with greater scientific recognition of its benefits, has been adopted to deal with the stresses of modern life. What is now referred to as the 'science of mindfulness' has been institutionalized in academic courses and programmes. Textbooks have been written for clinicians, mental health practitioners and mindfulness trainers. The marriage between neuroscience and mindfulness was described in a 2012 issue of the journal *Social Cognitive and Affective Neuroscience*: 'Mindfulness neuroscience is a new interdisciplinary field of mindfulness practice and neuroscientific research; it applies neuro-imaging techniques, physiological measures and behavioural tests to explore the underlying mechanisms of different types, stages and states of mindfulness practice over the lifespan' (Tang and Posner, 2012).

The brain and neural mechanisms involved in meditation or mindfulness training are the focus of the mindfulness neuroscientist. For the practitioner, mindfulness provides a drug-free method for dealing with the stresses of modern life (Williams, 2011) which can also potentially be piggybacked onto everyday activities, such as commuting (Davidson and Dahl, 2018: 62), and therefore adapted to a busy life. Perhaps most importantly, evidence suggests that mindfulness works. Meditation can change brain and immune function in positive ways (Davidson et al, 2003; Goleman, 2003; Doidge, 2007; Siegel, 2007; Gilbert, 2009).

In response to its proliferation, a critical debate has emerged regarding the origins of mindfulness in Buddhist meditation and concerns about distortions that arise from its introduction into a Western context. The criticisms are somewhat different depending on whether the focus is on the science or the practice. Placed in the historical context of Eastern meditation, a distinction between the study of mindfulness – that is, the science – and the practice is already problematic. In the Eastern traditions science and practice cannot be separated as they are in the West. Buddhist meditation has its roots in a much older science, originating with Hinduism, going back to the first millennium BCE. The methods of this 'supreme science' (*Brahmavidya*) were concerned less with the external world than with knowledge of the reality that might be discovered through the mind. The rishis or seers of ancient India identified, beneath the world of change, an infinite changeless reality that is said to exist at the core of every human personality (Easwaren, [1985] 2007a: 17). Vipassana, a Pali term, and vipasyana, its Sanskrit equivalent, refer to one of India's most ancient techniques of meditation, which was rediscovered by the Buddha. The term means 'special, super seeing', as is also reflected in the Chinese translation of Vipassana (*neiguan*, 内觀),[2] which translates into English as 'seeing'.

To see is to know, and 'knowing is impossible without seeing' (Suzuki, 1963: 235). Seeing for the Eastern meditator refers to a mode of perception that may include visual perception but also becomes a non-sensory experience of reality. As Capra ([1975] 1991: 43) notes, when the meditator talks about seeing, looking or observing, they emphasize the empirical character of their knowledge and the preparation of the mind for an immediate and non-conceptual awareness of reality. The techniques make it possible to silence the narrative activity of the mind and shift awareness to more intuitive modes of consciousness. The experience of oneness with one's environment is the main characteristic of this meditative state. Every expression of fragmentation and difference fades away into an undifferentiated unity (Capra, [1975] 1991: 47).

The ancient seers believed that the single most important purpose of life is to experience the infinite changeless reality in order to realize compassion on earth. In the context of an ongoing dialogue with Western scientists, the Dalai Lama pointed to systematic forms of enquiry, going back millennia, by which Buddhists have sought to understand human consciousness, a subject which

2 The character in the text is a traditional symbol, as distinct from the more contemporary simplified figure 円觃.

had received little attention given the external and materialist focus of classical Western science. The participants in the dialogue agreed that Buddhism shares much in common with Western science. Both are committed to enquiry and investigation and probing beneath surface appearances, as well as to notions that certain levels of reality only become accessible through special techniques of investigation (Harrington, in Davidson and Harrington, 2002: 19). Both traditions have also sought to recognize and address human suffering. Having said this, they rest on fundamentally different assumptions about the means and ends of enquiry.

Western science has sought solutions to human suffering through objective knowledge and manipulation of the material world, with significant results – particularly in the area of medicine – but these successes are often morally ambiguous, as evident in the creation of biological, chemical and nuclear weapons of mass destruction. In Buddhism, knowledge of reality is associated with self-transformation, the liberation of practitioners and an underlying assumption that all life within the universe is connected. It assumes that the transformation of human suffering is not only possible but is an ethical imperative.

Neuroscientists and Buddhists share an internal focus on mind and consciousness, as distinct from an external emphasis on materiality that is characteristic of classical science. However, the exploration of mind undertaken by the ancient rishis or contemporary Buddhist monks is at odds with the third-party approach of neuroscientists, although the latter do attempt to correlate their 'objective' findings with the subjective reports of the meditators.[3] A fundamental difference, as Waldron (2017: 1) notes, is that the 'monks have been treated as objects whose minds are studied by scientists, rather than as subjects who study minds on their own.' Neuroscientists have in the process tended to neglect the Buddhist philosophy in which the practice is grounded. In the one, scientific inquiry and practice are separate. In the other, they are inseparable.

The field of neuroscience generally treats the brain as matter, and as a property of the individual, although an increasing number of scientists have moved in the direction of a conception of mind that is more compatible with quantum science as well as Buddhism. Already in the 1990s, the decade of the brain, Richard Restak (1994: 111–21) noted that brain research on consciousness had cast doubt on traditional ideas about 'the unity and indissolubility of our mental lives', and particularly the notion of a 'unified

[3] Over the last decade, significant work on the neuroscience of mindfulness has taken at the Waisman Laboratory for Brain Imaging and Behavior at the University of Wisconsin-Madison. The Dalai Lama helped to recruit Tibetan Buddhist monks to participate in research on the brain and mediation. See Davidson and Lutz (2008).

free-acting agent'. Valera et al ([1991] 2017) around the same time (in a book since republished) sought to go beyond the classical conception of neuroscience to engage with the relationship between the researchers' understanding and the practice of mindfulness. At stake is the underlying conception of self in the two traditions as well as the relationship between the scientist, the apparatus of investigation and the object of measurement. The Buddhist science of meditation brings the observer, the apparatus and the subject into a much closer interaction. The Buddhist meditator is the observer of his or her own internal processes and any measure of them. For the Buddhist, and the quantum physicist, the observer and the apparatus (the mind in the case of Buddhism) are inseparable and the measurement is impacted by the relationship. By contrast, in the more classical conception of neuroscience, the focus is the physical brain, which is measured by a scientist whose involvement, it is assumed, does not impact on the measurement.

At the core of the mindfulness debate is a question of whether secular or scientific mindfulness is indebted to Buddhism, or is fundamentally different, and whether the Buddhist concept of self could ever be reconciled with a Western focus on the individual, which is fundamental to contemporary notions of what it means to be secular and modern. As Wendt (2015: 12) notes, most social science, including constructivism, rests on Newtonian assumptions. Based on these assumptions, subjects or objects are in the first instance material, possess an essence and occupy local space. The neuroscientific focus on the individual brain as well as the proliferation of the practice of what has been referred to as 'McMindfulness', 'secular' or 'selfie' mindfulness or 'capitalist spiritualism' reinforces a notion of the self as intrinsic to the individual (Hsu. 2016; Repetti, 2016; Titmuss, 2016; Doran, 2018). The objective of mindfulness, from this perspective, is first and foremost self-improvement and stress reduction.

The 'neoliberal container self' contrasts with the Buddhist claim that the self is an illusion. The illusion begins with 'I', which is experienced as a core essence that is unchanging. Attachment to the impermanent self gives rise to greed, hate and ignorance (the three poisons), which then often lead to reactions that harm others. The Four Truths of Buddhism rest on an assumption that all phenomena are in a continuous state of flux and change as well as lacking in any essence or essential self (*anātman*). The lack of an unchanging core self is a source of suffering and dissatisfaction because we want security and well-being, which we can never as living organisms achieve given our dependence on all sorts of conditions (Waldron, 2017: 2), and not least our dependence on others. If science seeks an objective understanding of how the world works, Buddhism instead prioritizes how we as sentient beings experience the world, and in particular our suffering and the potential for liberation from it. Dissatisfaction arises from the disjunction between the desire for security and the conditions of a reality where there is no essence to hang on to. It is, however, this search for security that leads to suffering.[4]

In contrast to the Buddhist emphasis on suffering, mindfulness practice in the US, critics argue, has been commodified and, as a big business, is worth more than one billion dollars in the US alone (Doran, 2018). The practice of mindfulness is itself a commodity that is sold to individuals. Other commodities add to the profitability, from expensive meditation gear, Dharmic dating services, Dharmic dentists and accountants and any number of commercial activities related to mindfulness (Sharf in Rosenbaum and Magid, 2016: 149), not to mention the proliferation of mindfulness apps and the one hundred thousand–some books that have been published on the topic. The commodification and creation of desire for a good that can be bought and sold with the objective of achieving a short-term alteration of mood or stress is, the critics argue, too far from the Buddhist emphasis on human suffering that arises first and foremost from attachment to self. What is instead required, it is suggested, is a deep change in one's attitude to life as a whole, allowing for a greater acceptance of suffering and death (Schneider, 2017: 773). The objective is less to rid the self of suffering than to change one's attitude toward it (Schneider, 2017: 783), which includes an ability to remain calm in its midst.

The individual brain is an object of study, which gives rise to conclusions that a particular practice that modifies the brain will result in the self-improvement of individuals. Individuals become consumers of a practice and of objects that will facilitate the maximization of self. The focus remains at the level of individual parts and the isolation of and focus on individual mental events, thereby detaching the practice from a more holistic perspective (Schneider, 2017: 782). Indeed, a focus on the individual brain, as distinct from the larger context within which the self is located, is a part of the problem: it confuses the biological conditions that make mindfulness possible with mindfulness itself (Thompson, 2017: 1). Cognitive neuroscience treats mindfulness as the inner observation of a private realm of thought, feelings and bodily sensations, but it should instead be understood as an embodied phenomenon, involving not only the brain but the body and the physical, social and cultural environment (Thompson, 2017: 1). Secular mindfulness fails to engage in deep reflection on the larger structural forces that shape its environment (Hsu, 2016: 369). Western secular thought and commodification focus on the isolation of parts, whether the individual brain or the individual subject or object, and their fixity in time and space. Buddhist philosophy, by contrast, starts with the whole, whether in the context of the individual or the larger universe.

4 The Buddhist argument shares a family resemblance with those of critical security theorists that searching for a perfect security, which can never be attained, is not only futile but counterproductive.

Another criticism is less about how mindfulness is applied in the West than the consequences of Buddhist detachment for action in the world; that is, detachment from self and world pacifies individuals and thereby does not contribute to change. Ng (2016) claims that, while helping to manage stress, therapeutic mindfulness deflects attention away from the very systemic and structural conditions that are a source of stress in the first place. Zizek (2001, 2012) argues that mindfulness is compatible with the interests of corporations. The enhanced ability of the individual to manage stress mindfully makes them more efficient and productive workers who contribute to the profits of the company. In the military context, mindfulness makes soldiers more resilient and capable of managing their emotions, thereby facilitating their return to the battlefield rather than leaving them incapacitated by trauma. The concern of the critics is that corporations, which are driven by profit, or institutions such as the military show little concern for the well-being of employees (beyond mindfulness training) or the societies from which they arise. Some corporations that use mindfulness training – for instance, Amazon and Google – have been notorious for tax evasion (Titmuss, 2016: 185, 189). Mindfulness, it is argued, not only stands in the way of change but contributes to the reproduction of the very conditions that create stress and the need for mindfulness in the first place.

Further criticisms have shaped the mindfulness debate. I have highlighted three of particular concern, relating to the underlying concept of self, which is commodified, the absence of attention to context and the assumed pacification that arises from detachment. Critics within the debate highlight the disciplining influence of a neoliberal concept of self and the limits it imposes on agency. A core claim is that far from moving away from the neoliberal conception, mindfulness serves to reinforce it and the institutions, both domestic and global, upon which modern culture is built. Mindfulness in the US thus rests on assumptions of self that are contrary to those embedded in Asian practices of meditation.

Turning toward the relational self

Critics argue that the practice of mindfulness in the West has been put to use for purposes of profit and greed, which, from a Buddhist perspective, is the very thing that gives rise to human suffering. Far from bringing about change, and a reduction of suffering, mindfulness makes soldiers or workers more resilient and thus able to do their job more efficiently, thereby reinforcing unequal global capitalist power structures. Those who would defend the practice as applied in a Western context start with a distinction between mindlessness and mindfulness. The central claim is that most of us go about our day-to-day life from a position of *mindlessness* and deal with stress through artificial intoxicants of various kinds. The defenders argue that *mindfulness* is a better way to deal with forms of distress that might otherwise lead to alcohol, drug or other forms of addiction. Mindfulness provides an alternative means to calm the mind and thus reduce stress, which may, with a continuation of the practice, lead to deeper change.

Repetti (2016: 473) defines mindlessness as not paying attention to what one is doing, thinking, perceiving and experiencing. It rests on an absence of awareness of the self. There are degrees of mindfulness. Anyone, he argues, whether engaging in mindfulness for purposes of ego or spiritual development, is likely to experience some mental freedom and this is inherently good (2016: 475). The idea that anyone can benefit is consistent with Buddha-nature theory, according to which all are naturally endowed with the Buddha potential within, although few will realize it.[5] The Dalai Lama, the primary representative of Tibetan Buddhism, is said to have asked Jan Kabat-Zinn, a major influence behind the US mindfulness movement, to develop a more secular form of meditation that would potentially increase the happiness of suffering Westerners (Repetti, 2016: 475).

The idea that any cultivation of mindfulness may give rise to more altruistic tendencies is demonstrated in the parable of the Buddhist thief, which comes from the philosopher Nāgārjuna, who will be discussed in more depth in the next section. In this parable, a thief asks to be a Buddhist disciple but says he wants to continue being a thief. The Master, Nāgārjuna, agrees to take him on but asks him to be mindful of his actions while stealing. Three weeks later he returns and says he can't do both because awareness of the implications of his actions gave rise to compassion for his victims which then prevented him from stealing. By extension mindfulness, in whatever form, argue the defenders, stands a greater chance of changing individuals and thus the world than mindlessness (Repetti, 2016: 482). For instance, mindfulness within the military will reduce the incidence of mindless acts of violence, triggered by emotions, that cause greater suffering for others and the self.

In contrast to arguments that mindfulness is purely for selfish individual ends, the thief example suggests several layers of a potential reorientation of self towards the world, which begins with the intention to start on a different path. The new awareness of self and the implications of its action for others leads to ethical reflection. The thief parable, while used to justify secular mindfulness also, when broken down in this way reveals the very different

[5] Buddha-nature theory is primarily associated with East Asian Buddhism, including various schools in the Mahayana tradition from China, Japan, Vietnam and Korea, which largely abandoned the more analytical and critical dimensions of the Indian Mahayana tradition (see Weitsman in Rosenbaum and Magid, 2016: 150).

objectives of Western science – that is, measurement – and Buddhism – that is, ethical reorientation.

The question of whether the Buddha exists within everyone or whether enlightenment is only available to the few is a debate within Buddhism itself. The point here is that the transition from mindlessness to mindfulness is first and foremost a *conceptual* transformation. The process starts with a concept of 'I' as separate and egoistic, arising from a narrative that is preoccupied with self yet lacking in self-awareness. Over time, the practice of mindfulness or meditation gives rise to consciousness of how one's own actions impact on others, and thus awareness of entanglement with them. Jon Kabat-Zinn (2005) refers to this as an 'Orthogonal Rotation of Consciousness'. 'Orthogonal', a term originating in Euclidean geometry, refers in this case to a rotation by which conventional reality is situated in a much larger three-dimensional space.[6] Such a rotation moves away from what I or we need to be secure, to ask different questions. As Jon Kabat-Zinn (2005: 360) notes:

> Just by asking, for instance, 'Who is suffering?' 'Who doesn't want what is happening to be happening?' 'Who is frightened?' 'Who is feeling insecure or unwanted, or lost?' or 'What am I?' we are initiating nothing less than a rotation in consciousness into another 'dimension,' which is orthogonal to conventional reality, and thus, able to pertain at the same time as the more conventional one because you have simply 'added more space'.

In this rotation, the world becomes a larger place which, viewed from another angle, is cast in a new light. The narrow view of self-interest is replaced by more compassionate possibilities which were inconceivable within the prior one-dimensional space.

The monologue of the intrinsic self gives way to a dialogue that transcends inside and outside. The inner narratives lose some of their status as ontologically real, that is, as a reflection of the world 'as it is'. The inner dialogue forces an engagement with, rather than destruction of, the tension and difference between narratives, both inside and outside. The first transition arises from a narrative of the egoistic self, with intrinsic properties. Narratives of self are embedded and entangled within a larger context and world. The egoistic self can, with this transition, be seen from the perspective of the relational self. The second is a transition from treating the narratives in one's head as ontologically real – that is, as a reflection of the world as it is – to treating them as both real and non-real and thus capable of being detached from self.

6 Orthogonal means intersection or lying at right angles or perpendiculars.

The rotation gives rise to a different practice of self that rests on an awareness of entanglement of the self in language. To refer to the distinction between the two in terms of West and non-West, or science and Buddhism, is perhaps too bold. There are Western thinkers, such as David Hume ([1739–40] 1964), or more recently Wittgenstein (1958), who would have opposed the idea of a unique 'I' or 'me' that exists as an intrinsic entity entirely separate from the world it inhabits. Hume (1739: vi) famously stated that 'I can never catch myself at any time without a perception, and never can observe anything but the perception.' Numerous systems of thought identified with the West, from that of Heraclitus to post-structuralism to Marxism, rely in different ways on a notion of self as a subjectively constructed narrative (Chappell, 2005: 220). That Buddha did so as early as the fifth century BCE is perhaps not in itself a rationale for focusing on the Buddhist argument, but the concern of this book with conceptions of self and mind within Eastern philosophy is. As suggested earlier, in Buddhism the intellectual construct cannot be neatly separated from practice. The Buddha was not first and foremost concerned with intellectual argument but rather practical engagement with a world of suffering.

The key issue raised by mindfulness or meditation regards the positioning of the self vis-à-vis the continuous stream of words that run through the mind. In the case of mindlessness, we are unaware of the words, treating our thoughts as arising from an intrinsic 'container' self and as a reflection of our essence and the world 'as it is'. Mindfulness is an attempt to take some distance from the narratives, watching them pass without accepting their status as true or ontologically real. The narratives are just allowed to be, without a need to act on them. Stepping back into the underlying philosophical assumptions helps to clarify why detachment from the narratives and the narrative self is less about inaction or an inability to act than transforming the relationship between the self who acts and the world of engagement.

Impermanence and motion

The mindfulness debate highlights the contrast between a focus on individual and brain in a society driven by desire, on the one hand, and, on the other hand, the illusory nature of the self in a universe of impermanence and change. The purpose of what follows is to deepen the analysis of what this debate implies about the nature of 'I' and the world or universe it occupies. The central question is whether the self, like any subject or object, can be said to exist independently of other selves or objects? Does the self have an unchanging core and essence? The Buddhist answer is no. We do all possess a sense of 'I', but this self is impermanent, changing and changeable and often driven mindlessly by a preoccupation with self-protection. The illusion arises from the human tendency to cling to some notion of a fixed essence and reality and to turn away from the acknowledgement of suffering and not least death.

By way of contrast, impermanence and motion can also be found in Hobbes' *Leviathan* ([1651] 1958).[7] In the field of international relations, Hobbesian self-interest is usually translated as pure egoism. However, political philosophers have over time debated the extent of Hobbes' egoism, from attributing to him the extreme position that humans are incapable of genuine benevolence to a more recent mainstream view that he acknowledges the potential for other regarding feeling (Slomp, 2018). Here I want to emphasize what he says about impermanence and motion and, in particular, e-motion, which is a term used by Sarah Ahmed (2004) to highlight the motion and outward movement of emotions. Hobbes identified two kinds of motion that are specific to animals, including the involuntary motions within the body that constitute life, from the coursing of the blood, the pulse or breathing, to those more voluntary motions involved in speaking or moving or in response to that which we see or hear which he associates with memory (Hobbes, [1651] 1958: 118). Bodily motions, and their manifestation in visible actions – walking, speaking, striking and so forth – are preceded by appetites and desires, starting with food and water, as well as aversions. Appetite and aversion, he states, 'signify the motions, one approaching and the other returning.' It is here that e-motions come in, with the love of that which is desired and the hatred for that of which aversion is experienced (Hobbes, [1651] 1958: 119). He states that the motion within us is caused by the action of 'externall objects' ([1651] 1958: 121), and refers to the simple passions of 'appetite, desire, love, aversion, hate, joy and grief' ([1651] 1958: 122). A concern with emotions is clearly evident in Hobbes in a way that has not, until recently been recognized in realist international relations (see Booth and Wheeler, 2009). In light of the longstanding marriage of Hobbes to materialism, the emotions have been parsed out.

Hobbes, like the Buddha, is concerned with the impermanence of self, as well as with desire and aversion as sources of suffering in the world. However, the Buddhist emphasis on calming the mind finds it opposite in Hobbes' claim that 'there is no such thing as perpetuall Tranquillity of mind, while we live here, because life itself is but motion and can never be without Desire, nor without Feare, no more than without sense' (Hobbes, [1651] 1958: 129). Discourse, he argues, extends not only outwards to the use of language in the social realm but to 'mental discourse', which 'consisteth

[7] It should be noted that Hobbes was still writing within the pre-modern tradition of the *élan vital*, or life force, which will be discussed in the next snapshot; however, he was also influenced by the mechanistic thinking emerging in his environment, which would be codified by Newton several decades later.

of thoughts that the thing will be and will not be or that it has been or has not been alternately' ([1651] 1958: 130) and a recognition that there can be no absolute knowledge of fact, past or to come ([1651] 1958: 131). The concern with motion, e-motion and the movement of language, in the mind and outside, contributes to Hobbes' conclusion that all humanity is inclined to 'a perpetuall and restlesse desire of power after power that ceaseth only in death' ([1651] 1958: 161). In the absence of a common power to 'keep them all in awe', humans will be in a condition of war, of 'every man against every man' ([1651] 1958: 185), within which life, with the 'continual fear, and danger of violent death', is 'solitary, poor, nasty, brutish and short.'

Hobbes was writing in the context of the English Civil War.[8] Realist international relations has built on the international implications of his argument. The focus of Buddhist mindfulness is more at the level of the individual, even while it rests on a philosophical system that questions the ontological status of the individual. The question is whether, in reading the Buddhist philosopher Nāgārjuna and Hobbes through each other, we gain insight into the potential implications of Buddhism for understanding global politics. Both start with a problem of motion and impermanence, which is the source of desire and aversion. Desire and aversion find expression in discourse. Discourse is internal to the thought processes of individuals and to the shared world of politics. The point of departure and the source of suffering in the world are not dissimilar. They are distinguished, however, by the conclusions that arise from this point of departure, although, here again, there is a similarity. In light of Hobbes' argument about language and motion, one might see the move towards a social contract as not only vesting power in an authority who will provide protection; this power is dependent on a narrative of the collective 'I'. The social contract determines which discourse or narrative of the collective self will prevail. So, for instance, with the conclusion of the Treaty of Westphalia, and against the backdrop of a brutally destructive war, the power to decide whether a narrative of Catholicism or Protestantism would underpin practice was vested in the sovereign prince. Or, in a more modern context, such as contemporary Britain, the proliferation of narratives about the future of UK/British/English 'self' that surrounded the Brexit vote was only resolved with the very clear majority vote for the Conservatives under the leadership of Boris Johnson in the December 2019 election. The latter rests on a more Lockean than Hobbesian conception of the social contract, but the result is similar, that is, the cessation of the proliferation of narratives with the accepted dominance of a single one which grounds the authority to further decide. Or so it seemed.

[8] The period of the English Civil War was 1642–51.

Hobbes' solution to the insecure self shifts the problem to another level. As individuals vest authority in a leader, the problem of 'war of all against all' is displaced onto the relationship between 'collective selves' or states. The literature within international relations has not sufficiently problematized the ontological move from the individual, capable of expressing emotions and concerned with bodily protection, to a body that lacks these physiological characteristics (see Wendt, 2004). While starting with the same problem, that is, motion and impermanence, the Hobbesian and Buddhist arguments give rise to very different conclusions about the nature of self.

Two truths

Notions of self that assume a separate and independent entity are indebted to classical physics and a materialist understanding of the world and universe. In this conception a physical and material body contains an ego or soul, and mind and body are separate. Given the focus on individual, body and ego, any move toward the collective is faced with the physical limits and causal closure of a universe composed purely of mindless matter, as expressed not least in the seeming solidity of the human body. The collective cannot be more than the sum of its parts.

Buddhism, like quantum physics, starts with a universe in motion, propelled by a dynamic relationship between emptiness and matter. Both replace a notion of matter as essence with a universe characterized by change and impermanence whereby becoming is more basic than being. While quantum physics focuses on impermanence at the subatomic level, Buddhism develops its implications for understanding the human self as both real and non-real.

Nāgārjuna, founder of the Madhyamaka (middle way) school of Mahayana Buddhism and the most influential and widely studied philosopher within that tradition (Garfield, 1994), set up a dialectic based on his theory of two truths, one of which is absolute and the other provisional. The absolute truth is that all things are empty and the provisional truth is that they exist, even though this existence is impermanent and fleeting (Soeng, 2004: 32). Emptiness, which is the centrepiece, is a reference to the absence of intrinsic nature, not in the sense of nihilism, as is often assumed, but in the sense that everything is fundamentally relational (Priest, 2009: 467). Nāgārjuna's concept of 'dependent origination' or 'dependent arising' highlights the extent to which everything, from objects to events, emerges from a complex web of interrelated causes and conditions, and that nothing exists by itself. There can be no whole without parts and without a whole there can be no concept of parts, and all phenomena lack an independent identity (Dalai Lama, 2000: 37–8). Because any thing is a compound entity, it has no core independent of the conditioning factors that are responsible for its creation. It is instead made up of a web of relationships that are dynamic in character (Soeng, 2004: 192). As Tao Jiang (2014) argues, Nāgārjuna is one of the few thinkers in either Western or Indian philosophy to avoid the tendency to separate the universal and the conventional, either by making them incommensurable or through their reconciliation, which makes the ultimate foundational and the conventional derivative. Nāgārjuna rejects such bifurcation into ultimate and conventional. Emptiness replaces any conception of ultimate reality as a background composed of substance or essence, yet this more radical conception of ultimate reality is intertwined with the conventional.

Within this conception, and in contrast to Hobbes, being is fundamentally relational and dependent on others. The identity of any one person has its origins in their birth to particular parents at a specific point in time, their having unique DNA, their going to a specific school, with certain friends, and their being affected by the things they do and see (Priest, 2009: 469) within a particular cultural context and time. According to Nāgārjuna, emptiness is a condition of interdependence, that is, that all things are empty means that all things are mutually dependent (Barnhart, 1994: 649). In this respect, dependent arising and emptiness are ultimately two words for the same thing (Sante, 2009: 135). There is no absolute, non-relational, independent presence that is unconditional (Barnhart, 1994: 652). Things are both separate and entangled, real and non-real. Ricard and Thuan (2001: 90) use the example of a tent to illustrate the point. If a tent is dismantled by separating its cloth, poles and rods, then the tent no longer exists, although the parts do. If the cloth is torn up, there are threads that can be reduced to fibres and further to molecules and then atoms and finally to particles whose mass is equivalent to intangible energy. The change from tent to particles, or the other way around, from particles to tent, contains, they argue, no discontinuity that would justify the distinction between microcosm and macrocosm.

An everyday conventional reality is produced from what is not real, or from the background independence of a universe that has no ontological primitives. Quantum physics explains the relationship between 'background independence' and emergent phenomena at the microscopic level, as discussed in the introduction. Nāgārjuna provides an understanding of how a universe that is empty of ontological primitives gives rise to life. For both quantum physics and Nāgārjuna, everything is real and not real, hovering between existential extremes until an observation is made. The physicist Laurent Nottale (1998:11) stated that:

> Some philosophers have … concluded that nothing including matter and mind intrinsically exists. If we trace the history of this line of thought back, it seems to have been first formulated in Oriental

> thought by Siddhartha Gautama (Buddha) over two thousand five hundred years ago. There is no nihilism in this concept, no denial of reality or existence, but rather a profound view of the very nature of existence. If things do not exist in absolute terms, but do nevertheless exist, then their nature must be sought in the relationships that bring them together. Only these relationships between objects exist and not the objects themselves. Objects are relationships.

The Dalai Lama (2005: 52) commented on the 'unmistakable resonance between Buddhist notions of emptiness and quantum physics. If on the quantum level, matter is revealed to be less solid and definable than it appears, then it seems to me that science is coming closer to the Buddhist contemplative insights of emptiness and interdependence.'

The illusory self

Dependent arising or origination provides a backdrop for understanding the self as empty of intrinsic identity. Subsequently, the notion that 'I' exist as an essence rests on an *illusion*. We do not exist in and of ourselves but only as a product of the relationships and encounters that have made us who we are, which, it should be reinforced, applies to all forms of life and not only humans. While often interpreted as a claim that we don't exist at all or that we are 'mere' language, dependent origination provides a richer conception that situates subjectivity firmly within a range of familial, geographical and cultural contexts which are conventional but nonetheless real. The main obstacle and source of suffering arises from a failure to see the narrative web within which the self and all its dependencies are spun for what it is; that is, as illusory and subject to change. We experience ourselves as autonomous actors with separate brains who react to our environment, including those who occupy it, as if entirely separate. Each of these separate beings is first and foremost concerned with protecting the container self.

The self is fuelled at the intersection of consciousness and language. The self provides direction, expressing the intention of the actor, while being dependent on a shared language in doing so. Conventional truth, or all of the various ways in which consciousness is made manifest in a relational world, arises from the process of producing 'I' and 'you', from the causes and conditions of the differentiation process, much as any object (for example the tent) is formed through a range of parts and conditions. Any 'I' is shaped over time and *changes within a relational web* within which potential meaning is super-positioned. Everything we see, hear, think is expressed and made manifest in the language available (the conventional apparatus) for seeing or perceiving in any particular context, and not least the language of 'I' and 'mine'.

For Hobbes there was no such thing as perpetual tranquility of mind or freedom from the swirling words and emotions. Nāgārjuna, by contrast, points to a potential to cease the proliferation of concepts in order to experience self and life from a different angle, not for the purposes of escaping or transcending reality but in order to engage with it from a different position. The different position recognizes the fundamental dependence of the self on causes and conditions that are larger than the self. The self is an *activity* and, as will be explored in the snapshots of Daoism, it is a navigator of a relational world rather than an independent object that is fundamentally at odds with it. To say that the self is empty is not to deny the physical body but to shift emphasis from an exclusive focus on materiality to the way that consciousness and experience mix in the constitution of particular kinds of subjects. The subject is a product of the relationships and practices by which it has formed and changed over time.

The change in perspective from one that sees parts as separate and independent to one that sees them as embedded with a whole is a theme that resonates with recent discoveries in general systems theory, evolutionary biology and cognitive science. Waldron (2002: 3) argues that the illusion of self is more than language but at the same time is inseparable from it. Consciousness, which is a focus of the next snapshot, is not a faculty of the brain that cognizes objects but rather a process that is dependent upon conditions. In place of a subject who observes an independently existing world, differences specific to context trigger consciousness arising from circular feedback systems. Learned linguistic categories, belonging to particular kinds of classificatory contexts, feed into the process of discerning and performing difference. Contrary to the rational actor who undergoes a rational procedure, the symbolic self draws on the reflexive potentials of language to make sense of their environment. The notion of a substantive causal agent, motivated internally by an essential self or externally (for example by God), is replaced by evolutionary processes of circular causality that give rise to forms of awareness and experience by means of which our 'world of experience' is continuously yet unconsciously mapped, clarified and constructed.

The key issue is less one of how 'I' act in the world than of how a particular relational pattern, which is dynamic, creates the conditions for certain practices to emerge (Waldron, 2002: 4). How does a particular relational web or context create a generative matrix of conditions by which a particular potential, say Y, emerges from conditions X. This suggests a structural coupling between language and world such that distinctions within the world are gradually built up on the distinctive structures of each living system through the entire history of an organization and its environmental interactions (Waldron, 2002: 210). The two commitments, to dependent origination and emptiness, highlight the primacy of practice, which avoids the need to identify a grounding in anything 'metaphysically "deeper" than the practices themselves' (Perret, 2002: 382). The ultimate truth of emptiness and the conventional truth of dependent origination are not mutually exclusive but rather two aspects of the same thing, despite the fact that they are apprehended by different epistemic instruments, appropriate to each (Garfield, 2010: 347). Thus, the ultimate truth of emptiness is non-conceptual, and meditation is a path to seeing, but conventional truth is bound up in concepts and words.

Thinking about these two truths as parallel to the particle and wave helps to clarify the significance of the relationship between form and formlessness. The particle and wave are mutually implicated, but the latter cannot be seen as the particle appears with the collapse into physical form. In Wendt's (2015) argument, the physical form is language. As Wittgenstein (1958) clarifies, there is nothing behind the language to which it refers. The same principle applies most radically to emptiness itself. It is not a void that exists independently behind the illusion, the ultimate essence of all things. Emptiness is itself empty (Garfield, 1994). Our only consciousness of emptiness is through the conventional form of language, and in this respect emptiness is an aspect of conventional reality. The convention of emptiness determines its reality (Garfield, 1994: 220) within the classificatory system of Buddhism. The convention provides the key to understanding the deep unity between the two truths.

As Nāgārjuna states in the *Mulamadhyamikakarika* (Chapter 24, Verse 18):

> Whatever is dependently co-arisen
> That is explained to be emptiness
> That, being a dependent designation
> Is itself the middle way

The point is not one of non-existence or being unable to act but that who we are and what we do is a product of all that has shaped us over time combined with the degree of consciousness each brings to this mix, and how these combine in context. Manifestations of consciousness at any one time and place will express the causes and conditions which have shaped it to that point in time and of the particular context within which the self acts. While there may be no ultimate foundation for our actions, we are all a product of the causes and conditions – that is, the dependent origination – from which we have emerged. Our dependent origination is always the backdrop against which we think, say and do, or rethink, resay and redo. Rather than an *intrinsic self* that is good or evil by nature, or damaged or whole by nature, this is a *reflexive self* who engages with the world from the position to which their dependent origination has brought them at any one point in time.

The Buddhist contrast between dependent arising and emptiness has resonance with critical constructivist or post-structuralist arguments about identity, but goes a step further. For the constructivist, identity and practice belong to a context (parents/children; teachers/students; soldiers/enemies) and the world is a conceptual world of our making. For post-structuralists, such as Butler and Derrida, as in Buddhism, the self is performative and nothing beyond the performance or practice itself. Dependent arising expresses a similar point about the role of language; however, it goes a step beyond the performance of self to a practice which seeks to look beyond the language of 'self', 'mine' and 'I' to cease the proliferation of concepts: that is, the practice of meditation and an experience of emptiness without words. The practice temporarily detaches the self from its conventional form in order to re-engage with it in a new way.

Ceasing the proliferation of concepts

The illusion of 'self' is conventional and conceptual. The distinction between 'I' and 'other' is formed in classificatory systems that belong to language. The relationship between the embodiment of this language, intention and consciousness will be revealed in the next snapshot. Here I want to explore how the language itself often disguises or obscures entanglement with others, even while the relational self, formed in language, is a product of the stories we tell ourselves and to each other over time. The narratives constitute the self as solid and separate, striving to protect itself and confirm its own solidity (Mitra and Greenberg, 2016: 415). The narrative self is characterized as an 'evaluative self' which reflects an autobiographical narrative that is reconstructed from the past and projected into the future (Vago and Silbersweig, 2012: 6). According to Dennett (1992: 418), the narrative self exists in the dimensions of time and history, and accompanies our experience of personal agency. 'We do not consciously and deliberately figure out what narratives to tell and how to tell them; like spiderwebs, our tales are spun but for the most part we don't spin them; they spin us.'

The narratives tend to become more rigid as they are conditioned over time through their repetition (Vago and Silbersweig, 2012: 20) and their reliance on language, episodes, autobiographical memory and imagination. Mindfulness is important for beginning to increase awareness of the fixed boundaries between self and others constructed in language, and thus for decreasing reliance on these boundaries (Mitra and Greenberg, 2016: 416). Constructions of self, which often rely on that which is not self, come to be taken for granted. The significant shift from mindlessness to mindfulness is from understanding self and other as being in a distinct, mutually exclusive relationship to seeing their mutual implication. To say that the self is both real and non-real is to acknowledge that it exists as an *experiential* phenomenon but that experience has been constituted in relation to others, and over time, such that the experience extends outwards to its dependence on others, and inwards to the conception of self. All suffering stems from dependence on others for existence, from physical nourishment to emotional well-being and intentional action (Mitra and Greenberg, 2016: 418). It is this entanglement and failure to distinguish the real from the not real that leads to mindless reactions, which then further solidify and reinforce suffering.

We are not individuals who have sprung from the ground like mushrooms, completely separate and fundamentally at war with all, as described by Hobbes. We are instead a product of everything and everyone who has shaped us. Far from nihilism, this suggests a radical relationality. The Buddhist practice of meditation is about peeling back the layers of illusion. In this respect, relationality, or being bound up with others, is not by definition positive. It represents attachments to and entanglements with people or things that can be a source of suffering, whether due to loss, desire or aversion. The important issue, which makes this a problem of language, is that the reason for detachment is not first and foremost to protect one's individual self from other humans or other sources of potential harm; it is rather to increase the porosity of the border between self and others, a border constructed around the activity of 'I', 'mine' and 'yours', such that compassion might become possible. The purpose of detachment is to resituate the (illusory) self vis-à-vis this suffering in order to engage with it in a different way, one more akin to a conversation (Lea et al, 2014: 61). The turn implies a repositioning of self and other that starts with the individual and percolates outwards. The individual repositioning has a potentially transformative impact on the larger context. In taking distance from the stream of narratives passing through the mind, the meditator becomes aware of the extent to which language is a product of the assumptions contained within it – the classificatory categories by which the world of any one self is constituted.

At one end of the spectrum, the self is experienced as an intrinsic Hobbesian container which, to survive, must protect itself against all others within a 'war of all against all'. At the far other end of the spectrum, the Buddhist self carries with it a sense of 'I', but recognizes the self as bound up in and capable of compassion for all living things, which then opens up a potential to suffer with them (*com*, meaning with, together, plus *pati* meaning to suffer, the etymological roots of compassion). The crucial point is that this is a spectrum, a process and a journey rather than a single choice. The emphasis is on micro-engagements with the world, from the thoughts running through one's mind to more concrete actions in the world which are embedded in a process of peeling back the layers of narration, which any one is capable of beginning but few actually achieve. It is an opening of the container self to its embodiment and embeddedness in an entangled world.

The tension between Nāgārjuna's two truths, the one ultimate and the other conventional, leads logically to a conclusion that no one conceptual construction can be elevated above all others as truth, including Buddhism, in any of its forms. Buddhism, like any other -ism, has multiple manifestations in the world, from Theravada to Mahayana, and further distinctions within each and within and across cultures. Buddhism is, in some accounts, a religion, and in others a philosophy, a metaphysics or a cosmology. It is both real and not real, like the self or any other conventional construction, including this argument. All of these manifestations are a product of distinctions drawn in language. According to the sceptic, we have now reached a point of relativism from which nothing can be meaningfully said or done, and thus a form of nihilism, akin to former UK prime minister Theresa May's claim that to be from everywhere is to be from nowhere. If we are unable to ground the self or its action in anything concrete, we are left without foundations.

Nāgārjuna's casting of the relationship between language and world would seem to counter the argument, within the mindfulness debate, that there is *a* religious Buddhism that is contrary to *the* modern Buddhism of America or the mass Buddhism that began with British colonialism. From the perspective of dependent origination and emptiness, we are all the product of different conventions, or as Hume might say, conversations (see Kratochwil, 2018), related to our particular arising. If America is a context of dependent origination that has been shaped by a set of causes and conditions, which has in the past evolved on the basis of assumptions quite opposite to the teachings of the Buddha, a reorientation from mindlessness to mindfulness is a small step in a different direction. That potentiality may or may not be realized, but does nonetheless, in any account of Buddhism, point away from the egoistic self-maximizing that is associated with modernity or capitalism. Mindfulness, which has its roots in meditation, is potentially about developing an ability to 'see' in this way, but the process involves a transition between two distinct language games and, subsequently, experiences of self. The Buddha's story is about a search for meaning by a self that progresses through several stages, starting from an inability to see suffering in the world, to a recognition of embeddedness in a world of suffering which paradoxically opens the way to liberation from it.

The collective self

From a Buddhist perspective, or indeed that of Hobbes on closer reading, the individual self and the collective self are not qualitatively different but rather a function of the stories that we tell ourselves, as individuals or societies. It is the authority vested in narrative, and the activity of 'I'-ing, that is the point of departure for action rather than a self with an intrinsic identity. Some reflection on the nature of the collective 'I' is useful, not least because it more obviously rests on an illusion than the individual 'I', which is a product of direct bodily and conscious experience. Despite the obvious illusion, states act meaningfully in the world, and it is the larger narrative of the collective 'I' that provides the point of departure for the practices of the institutions by which states are constituted. This argument shares a family resemblance with a strand of literature that focuses on biographical narratives of the state. Berenskötter (2014), for instance, conceptualizes the nation state as a bounded community that is constituted by a biographic narrative that situates the collective in space and time and thereby constitutes a sense of community. The use of 'biographical' can be distinguished from the autobiographical in so far as the authorship lies outside the life experience to be accounted for; that is, a community has no single author who both writes and is the subject of experience. While the reading of the nation/state as a narrative already has a place in the constructivist literature (see, for example, Bially Mattern, 2005; Hansen, 2006; Steele, 2008; Malkso, 2009), Berenskötter brings to the table a specific notion of biographical narrative which he develops from Heidegger's ontology of being in the world. In this argument, the nation/state is an entity that comes into being through a narrative that designates a space of experience, which gives meaning to the past while also giving meaning to a horizon of experience and possibility in the future (Berenskötter, 2014: 264).

From this perspective, the world is constituted from multiple overlapping biographical narratives of separate states. Much like the narrative self at the individual level, the biographical narrative of states contributes to the configuration of an 'I' or a structure of knowledge private to a community by which a phenomenon of experience is delineated. The community comes to know things, including a collective 'self', through experience. As experience is ongoing, this places emphasis on the process of coming into being, a process which is always incomplete (Berenskötter, 2014: 268) and that is similar at the individual and collective level. As Ringmar (1996: 452) suggests, just as individuals come into being through the stories told by and about them, the narratives through which the state is cast as a meaningful entity bring it into being. Bially Mattern (2005) argues that the state narrative constitutes a form of consciousness regarding the relationship between self and world. This self is not merely a 'rational actor' but also consists of emotional and moral components (Steele, 2008: f10, f71) which come together in the activity of self.

The Buddhist concept of dependent origination grounds the narrative self in a universe of form and emptiness, thereby clarifying the nature of relationality and the entanglement of the illusory 'I', whether individual or collective. The world is constituted through multiple overlapping biographical narratives of 'separate' states, each of which, as sovereign, appropriates a sense of the 'I' who acts. If the narrative self/state emerges out of experience, it is both real and non-real. It is real in that it expresses the sedimented knowledge of embodied people and institutions. Having said this, any one self, state or individual could have developed otherwise, and indeed a global experience characterized by multiplicity reinforces the extent to which the collective 'I' *has* emerged differently in different times and spaces. From this perspective, the collective 'I' is both real and non-real, entangled with experience yet lacking in ultimate foundations. This 'I' has a certain intractability due to the sedimentation of habit, but is also malleable, given that the process of becoming is ongoing. While biographical narrative and dependent origination are similar, the former is less able to account for the entanglement with emptiness and thus the potential relevance of something like mindfulness for the political or international realm, or for expressions of a collective 'I' beyond the nation state. Indeed, the cessation of narrative at this level might on first appearance seem contrary to any notion of democracy, and in a global capitalist economy, subject to the same pressures for commodification and profit that characterize the Western mindfulness debate at the individual level.

The Brexit debate regarding the UK's departure from the European Union provides some insight into what dependent origination and emptiness, and by extension mindfulness, might add to the conceptualization of collective self. Individual and societal narratives are interwoven. Both are entangled in a web of words and argumentation. In the context of the Brexit debate, the multiple narratives pulled at the collective self, making any position of sovereign agency all but impossible. This was precisely Boris Johnson's point in arguing that an obstreperous parliament was standing in the way of action and thereby fostering uncertainty. The landslide election of the Conservatives in 2019 under a banner of 'Get Brexit done' made it possible to exercise collective agency.

The Conservative point of departure for Brexit was an argument that the UK needed to 'take back control' and reclaim its sovereignty. The UK performed as a sovereign entity with an intrinsic identity that was capable of acting with intention, a self-conception that had at the very least been muddied, and at worst abandoned, as a result of membership in the European Union. This intention, it is said, was expressed through the democratic will of the people in June 2016 to withdraw from the EU. The process that followed revealed the illusion that underpins the claim, raising fundamental questions about the collective 'I' of action. Far from a singular intentional entity, the process played itself out in a conceptual war of words, fuelled by social media.

The single narrative of 'take back control' disguises the dependent origination of the UK's self from multiple layers of experience and the extent to which the British sovereign self of the nation state is bound up in narratives of empire which belong to a particular historical context (Bell and Vucetic, 2019; Saunders, 2020). Any contemporary expression of a sovereign UK self is different than that of, for example, the British Empire. In the changing configuration of the world, previous colonial subjects, such as India, have emerged as major powers on the global stage. Further, the attempt to construct a sovereign UK separate from the EU carried the threat of a constitutional crisis as constituent parts, such as Scotland, sought to break away to create a separate national 'I'. At the level of historical experience, the potentials pulled in multiple directions, yet the contemporary context was very different. All is changing and impermanent. While the Hobbesian framework reinforces the outcome – sovereign decision – it also reveals the instability of that outcome, precisely because it relies on the constitution of an 'I' defined by its separation and difference from others, which was still contested from within. Buddhism highlights the illusory nature of the sovereign 'I' and asks how the relational self might differ.

What is the route out of this conundrum from the perspective of dependent origination and emptiness? Meditation isn't really an option at the international level. Edkins (2002), in her discussion of the trauma following the attacks on the US World Trade Centre and Pentagon on 11 September 2001, suggested the need to let narratives circulate rather than closing down on a single one. However, allowing multiple narratives to circulate requires a particular stance of detachment and mutual respect, which was largely absent from the Brexit debate. It is less the selection of *a* narrative or the circulation of narratives that is key than the acknowledgement of their status *as* narratives (as opposed to representations of reality 'as it is'). The acknowledgement then makes possible a reorientation. While collective meditation is not an option, a reorientation towards multiple narratives gives rise to different kinds of questions, as suggested by Kabat-Zinn (2005), expanding the focus from self to a relational world and questions of who is suffering and in need of care.

An orthogonal rotation arises from an acknowledgement that all life is suffering. Suffering arises from a search for security in conditions where all is in flux. Such a rotation might give rise to a recognition of the feedback loop and the way in which constructions of self and other – initially, in the case of the UK, in relation to refugees/migrants and the EU – boomeranged back, deepening divisions and hate within society. The COVID-19 pandemic further exposed layers of suffering, not least of the successor generations of historical victims of empire, who were more susceptible to the virus because of ongoing structural injustice or their position on the frontline. In the lead-up to Brexit, the suffering of other parts of British society – historically in the context of class structure and, over the previous decade, in the context of austerity – was blamed on the EU, throwing up the bizarre phenomenon that a significant portion of the working class voted for the Conservative Party, which embodies the other side of an historical hierarchy. The circulation of 'fake news' through social media heightened the proliferation of narratives.

The Brexit debate revolved around a fragmentation of narratives of the collective self, indebted to different points in its history: The Great Britain of empire imposed a narrative on the world; the Britain of sovereignty was an appropriation of self and 'mine' which sought to protect itself from the vagaries of conflict on the continent; the Britain of partnership in the EU relied on a much different construction and appropriation which sought to embed itself in a 'we'; the Scottish 'I' existed in the tension between its history as a part of Britain, including its role in the British Empire, and its own historical loss of sovereignty, combined with aspirations for independence. All of these narratives have been constituent parts of the British self, and, in competition, threatened to pull the country apart. The resolution in an election was an appropriation of the sovereign 'I', with echoes of empire in the background, which made it possible to proceed with withdrawal from the EU but portended further divisions in the making.

Conclusion

Buddhism provides an account of the emotional problem of consciousness and life in conditions of impermanence, thereby revealing the more human implications of an uncertain quantum world. It exposes a distinction between a universe composed of material form and essence (classical) and one of continuous movement and change (quantum). These two conceptions, as they relate to self and mind, came head to head in the mindfulness debate in the US. Critics argued that the practice of mindfulness in the US reinforced an individual, neoliberal concept of self that is contrary to Buddhist non-self and relationality. Proponents argued that any engagement with mindfulness practice creates a potential for transformation.

The problem of the illusory and relational self was further unpacked through a contrast between Hobbes and Nāgārjuna, both of whom start with the problem of impermanence but come to different conclusions. In one interpretation of Hobbes, the proliferation of concepts and the separateness of self are managed through the imposition of a social contract which displaces the problem of motion onto the collective relationships between states. In Buddhism, the cessation of the proliferation of concepts gives rise to consciousness of suffering and the impermanence of life. The latter creates the potential for a relational reorientation that recognizes the fundamental interdependence and unity of all life. The collective implications of this distinction were further revealed in the contrast between multiple narratives in the context of debates over Britain's withdrawal from the EU.

The UK does not have a singular identity that extends unproblematically, as a unitary essence, from the Magna Carta to the present. It is a much more messy, entangled and impermanent phenomenon, constructed on layers of suffering that remain largely unseen. The election of the Conservatives in

December of 2019 seemed to represent a new social contract. The events that followed in the spring of 2020, with the spread of COVID-19, led to an imposed calm, and cessation of activity, as societies across the globe went into lockdown. The imposed calm seemed initially to reinforce the sovereignty narrative, as the outpouring of compassion gave rise to a concept of 'we' within the UK. With time, the layers of human suffering disguised by the illusion of 'I' and 'we' began to surface, revealing the historical entanglement between the British 'I' of empire and its contemporary global legacy. The significance of these unfoldings will be explored in the snapshots to come. The next snapshot focuses in on consciousness and mind as they relate to action.

SNAPSHOT 2

Mind/No-Mind

Ceasing the proliferation of concepts is one objective of Buddhist meditation. Such a cessation is not purely mysticism; it is bound up with action in the world. The latter point is evident in the historical relationship between practices of meditation and war. Yoga, while not itself related to fighting, includes postures and breathing techniques that have been incorporated into the Indian martial arts (Kumar, 2017), both of which draw from the Vedas and are formed by the same basic principles. There is a long history of Buddhist monks who took up arms, including the warrior Shaolin monks, the Zen Buddhist warrior monks in Japan, who influenced Bushido culture, and in contemporary Southeast Asia, monks in the Theravada tradition (see, for instance, Jerryson and Juergensmeyer, 2010; Lehr, 2019). Meditation has an important place in ancient Daoist military practice (Meyer, 2012), and Tai Chi, a form of moving meditation, has its roots in the martial arts with foundations in Daoist teaching. As will be explored in Snapshot 4, ancient military strategy in China makes reference to practices of meditation in the context of battle. The historical relationship between the practice of war and of meditation is not obvious, but rests on a complex and paradoxical relationship between mind and ethics.

Meditation, as an ancient science of mind, seeks a greater understanding of an underlying reality and the realization of compassion on earth. Killing, except perhaps on compassionate grounds, would seem to be contrary to the latter objective. Nonviolence is at the heart of Buddhism in particular. The Buddhist warrior would thus appear to be a contradiction in terms. Within Buddhism, as Demieville (Jerryson and Juergensmeyer, 2010: 18) notes, 'nothing is more worthy than not-killing.' Murder is the first and most serious of ten major sins on what is referred to as 'negative paths of karma'. Killing is a part of warfare.

Within Mahayana Buddhism in particular, murder, suicide, self-sacrifice or warfare can be justified in exceptional circumstances so long as carried out with the right intention and without generating a harmful attitude. A conception of mind is at the centre of this ethical paradox. One reason for forbidding violence is that it is harmful to the actor, giving rise to a state of mind that the practitioner of meditation seeks to overcome (Maher in Jerryson and Juergensmey, 2010: 85). Killing that is legitimate must arise from right intention. Right intention is a state of mind that is attentive to suffering in the world, and arises from compassion.

Vitale (2014) argues that compassion, particularly within Mahayana Buddhism, builds on a particular logic: We only see suffering in the world as our own if we carry a discrete notion of self. Buddhist practice seeks to deconstruct the self, at which point the suffering of others becomes one's own suffering as well. Enlightenment is thus intertwined with the suffering of others. Sentient life, in this framework, is not exclusive to human beings. A life force infuses the entire universe, at the interface of the two truths of emptiness and dependent origination. All sentient beings are infused with this life force, which is extinguished as the body that houses it is destroyed. What had been a living, breathing being becomes a lump of matter. Death happens to all sentient beings. Natural death is consistent with Buddhist notions of impermanence. However, an intention to take the life of a sentient being is difficult to accommodate within this system of thought. A tradition that places the suffering of sentient life, and compassion for that suffering, at its very core would seem to have no place for killing.

This snapshot zooms in on the link between mind, action and ethics as regards the hard case of killing, particularly in a context of war. Emptiness, in this context, is a quality of mind that is the wellspring of action, particularly for Zen Buddhism, which is the subject of exploration. The concept of 'no-self' (*muga)* is often, in Zen texts, referred to as 'killing' the ego or one's own mind (Deshimaru, 1982: 51). The death as such is in the first instance a relationship to consciousness rather than the material body. Consciousness is a concept of quantum mind, which Wendt (2015) distinguishes from the classical emphasis on matter. The combination of no-self and 'no-mind' as it relates to action in war points to an agency that is agentless. The ethical paradox regards how, given the nonviolence at the core of Buddhism, this agentless agency could be compatible with an intention to do harm.

In what follows, the paradox of no-mind and war is read through a scene from a Western cultural production of the confrontation between the historical Japanese samurai warriors, who were said to practice Zen Buddhism, and the introduction of Western industrialized warfare to Japan in the nineteenth century. Based on a true story, the production brings a particular contestation between two types of warfare to the fore, while revealing a number of contrasts which may be overblown in the Hollywood production but nonetheless illustrate the relationship between emptiness and mind in battle, that is, no-mind. The film also addresses the relationship between violence and trauma, which is useful in connecting to the final section of the paper. As I will discuss, it is not only the film but the samurai as a historical phenomenon that are a cultural production. Casting the characters in a theatrical context resonates with the illusory nature of the self and conventional reality, as discussed in Snapshot 1, highlighting the notion that they are both real and non-real. The self is real as a relational phenomenon within which mind, body and meaning are woven together; it is non-real in so far as it lacks an essence. Just as the self arises from a relational context – in dependence on others – the same can be said of conventional truth. The analysis is not intended to be an historical investigation of the samurai but rather a philosophical one regarding the paradox of agentless agency, no-mind and a warrior culture that is embedded in a nonviolent tradition. The Western cultural (re)production provides a foil for clarifying what is at stake.

This snapshot connects self to mind and action while also revealing how something formless, that is, no-mind, can express intention and action, and how this action, embedded in a classificatory scheme that celebrates nonviolence, could be put to service by a warrior culture such that it is more than contradiction. The first part explores the debate surrounding agentless agency and its relationship to more familiar concepts of free will and determinism. The second shifts to an elaboration of the relationship between consciousness and sentience. The third then sets the stage for the cultural production and explores the working of no-mind within it. The fourth moves to an investigation of the ethical justification for an intention to do harm. The final section returns to the present, to explore a further paradox that arises from the use of techniques related to no-mind in the healing of US soldiers who suffered from PTSD as a result of their experiences of war in Iraq and Afghanistan. The problem of no-mind links to that of no-self, as explored in the previous snapshot, and no-action, which will be the subject of the next. As Zen Buddhism has been influenced by Daoism, it provides a particular angle and transition into the discussion of the Daoist concept of *wuwei* and the strategic thought of Sun Tzu.

Agentless agency

Critics of mindfulness, as explored in the previous snapshot, argue that it is a liberal technology of discipline that limits agency. A practice intended to create consciousness of and detachment from ourselves as egoistic individuals, when placed in a Western context, pulls back into assumptions of liberal theory. Far from creating agency, mindfulness, in this context, some argue, inspires passivity. Agency without an agent instead suggests a polarity that, like particle and wave, is mutually implicated even while contradictory. A similar concept of what Loy (1985: 79) refers to as non-dual action exists in all three traditions, but each provides a somewhat different snapshot of the same phenomena, focusing on a constituent part of the assemblage that constitutes action. Zen Buddhism highlights the problem of no-mind and action that arises from no-self. Daoism, as will be explored in the next two snapshots, provides a framework for thinking about the nature of action that flows from the illusory 'I' which is commonly referred to as *wuwei* or 'actionless action'. The concept of karma, which has a somewhat different import in Hinduism and Buddhism, provides insight into the relationship between 'fruits' of action and entangled time. Each separate snapshot represents a different angle on the phenomenon of action.

The question of how agency arises from a self that is lacking in essence is the focus of a literature that intersects with the mindfulness debate and explores the problem in relation to concepts of free will and determinism. Some would argue that the free will problem is a Western cultural and religious artefact that doesn't arise in Buddhism (Gomez, 1975; Harvey, 2007). The problem of agency relies on a metaphysics that assumes the existence of a will and an ability to direct our actions, choosing between alternatives and acting for reasons, as distinct from behaviour that is caused by external events. Any sense that we are determined by the latter would then undermine agency. It is for this reason that I have avoided use of the term 'agency' in the larger context of this book, focusing instead on an assemblage of mind, action and strategy.

In the tradition of Aristotle through to Locke, actions that are determined by our intentions are understood to be free, but causes that do not arise from the will are not (Repetti, 2019: 47). In this respect, freedom is not the absence of determinism but the presence of self-determination (Repetti, 2019: 48). Agency is a causal force of mind in the world which is distinguished from the causal force of the world on the mind. In the West, it is freedom that makes us persons, and responsibility for our actions arises from this freedom. The social contract tradition, from Hobbes to Locke and Kant and its contemporary descendants, treats the 'subject' as a self-contained, unencumbered, rational and a priori entity who performs a voluntary act of political contract (Williams, 2002: 24). The liberal conception of freedom and free will depend on an ontology of human essence. Freedom is the conscious and voluntary potential of the subject to act without external impediment.

Beyond free will and determinism

In so far as something resembling free will and determinism exists in Buddhism it looks quite different, for several reasons. First, the Buddhist concept of self is embedded in a different cluster of concepts. The liberal concept of self is surrounded by questions regarding the degree of the self's autonomy and the extent to which the self is determined in its action by causes external to it, which suggests an ontological separation between inside and outside. The metaphysical notion of a self with a soul, which is the basis for formulating the free-will problem, does not exist in Buddhism (Garfield in Repetti, 2019: 48–50). Buddhist emptiness of self is an expression of relationality and interdependence with others rather than of a separate soul. The self is intertwined with context and environment, including those who populate it. In so far as free will assumes autonomous human agency and a unique self, it would be one of the many illusions that humans invoke in the attempt to escape the suffering that is a part of life. This self, designed to protect us from suffering and awareness of impermanence, is at one and the same time our greatest defence mechanism and our prison (Brazier, 2003: 32). The notion of liberation that is implicit in the Third and Fourth Noble Truths provides a way out of suffering and a human capacity to escape self-delusion through adopting the noble Eightfold Path. As Batchelor (1998: 10) notes, there is 'nothing particularly religious or spiritual about this path' in so far as it 'encompasses everything we do' as an 'authentic way of being in the world.' Indeed, mindfulness contains a potential freedom to stop and reflect on things before acting and the ability to examine emotions and motives and direct how these influence actions as one engages with the 'interacting dance of rapidly changing mental states' (Harvey, 2007: 84).[1] In this respect, the paradox of no-self generates a further contradiction: we become more self-aware and able to act consciously once the self is recognized as an illusion.

Second, determinism, which is contrasted with free will in Liberal thought, is contrary to the presumed indeterminism of the world and universe in both Buddhism and quantum physics. The emphasis on impermanence and change in quantum physics or Buddhism stands in stark contrast to the emphasis on determinism in Newtonian physics. Resituated within a world of impermanence, the problem is less whether choice exists than how the self navigates a context of radical uncertainty wherein the grounding of reason or ethics in laws of nature or first principles is questionable and, in any case, too cumbersome, given that the environment can rapidly change. J. Peter Burgess (2015) makes a similar point when he calls into question the search for certainty and stability that underpins concepts of security and ethics. He argues instead that ethics is about making decisions in conditions of uncertainty and in the absence of adequate knowledge of the future, precisely because a context offers multiple possible futures. Insecurity is the space where choice exists and thus the space of ethics, giving rise to a range of tools and actions that assume less than perfect knowledge of the future. As presented in the previous snapshot, the problem of impermanence

---

1 While both pre-Mahayana Pali and Mahayana Sanskrit sources identify some compatibility between free will and the absence of a real self, the Sanskrit is more inclined to indeterminism and the Pali to determinism (Repetti, 2012: 193).

and unceasing motion are evident in Hobbes' state of nature as well as in Buddhism. The Hobbesian social contract imposes a degree of certainty on an uncertain world, while also shifting the problem to the relationship between states in a condition of anarchy. The Buddhist answer to impermanence is to cease the proliferation of concepts and calm the mind in order to more effectively navigate the various streams of consciousness in relation to an uncertain world.

A third point is that Buddhism is defined more by a pragmatic concern than the philosophical one of free will and determinism. Buddhism seeks to reduce suffering, as articulated in the Four Noble Truths, rather than to provide an answer to the question of whether free will exists. As Govans (Repetti, 2019: 20) notes, the Buddha was only interested in discussing topics that relate directly to overcoming suffering. To engage in philosophical debate while the house is burning down around you misses the point, that is, the need to act. To focus on determinism and whether you are *able* to act is counterproductive to removing the obstacles to action. Having said this, Buddhism does incorporate a philosophical argument about the cause of suffering and the nature of the obstacles, as expressed in the second Noble Truth: suffering arises from attachments or cravings. The question of how to remove the obstacles then follows from this in the Third Noble Truth, that is, ending attachment will bring an end to suffering. The Fourth Noble Truth shows how to do this through the Noble Eightfold Path, and further concepts of right view, right resolve, right speech, right conduct, right livelihood, right effort, right mindfulness and right *samadhi* ('meditative absorption or union') (Vetter, 1988: 11–4). Action that is more or less conducive to bringing an end to suffering is thus possible. We are still responsible for our situation and destiny, as expressed in the karma doctrine. The purpose of this analysis is to emphasize the experiential implications of the Buddhist approach to mind and agency. Given that there is nothing beyond the practices themselves, the objective is to look at how the broader framework for understanding mind, action and strategy has been applied in practice. The specific meaning and implications of karma, which means action, will be the focus of section three.

Emptiness and dependent origination

Some have argued that the possibility of agency belongs only to the conventional reality of dependent origination. Karin Meyers (in Repetti, 2019), for instance, argues that the problem can be solved by appealing to the distinction between ultimate and conventional truth. The connection between agency and responsibility in the Madhyamaka, Meyers argues, only addresses dependent origination or conventional truth. Through aggregation, appropriation and recognition of a body, of thoughts, values, dispositions and intentions as *mine*, selves are constructed. Importantly, the narration of life cannot be separated from physical and cognitive processes, which are intertwined with social life through our common narration and construction of each other.

Action does take place within the conventional world. However, the designation of agency as *purely* conventional also confuses the issue, given that emptiness and dependent arising are inseparable. Action in a Buddhist framework is closely bound up with practices of meditation. The point here is that emptiness cannot be separated out from dependent origination and both are equally empty, that is, dependent on social and relational convention rather than essence. Nāgārjuna applied the concept and logic of emptiness to all phenomena, including emptiness itself. As Garfield (1994: 220) argues, if he were merely saying that all phenomena are empty, this might appear to give emptiness a separate status as the essence of all things. But this is different than the claim that emptiness is itself empty. Emptiness is an aspect of conventional reality rather than a void standing behind a veil of illusion represented by conventional reality. While there is a temptation to conclude that nothing really exists except as a formless void, the emptiness of emptiness establishes that the entire phenomenal world, including human beings, is recovered within that emptiness (Garfield, 1994: 238). However, it is recovered as dependent and relational rather than as existing in and of itself. Emptiness does not mean non-existence. The Madhyamaka positions self at the interface between the ultimate and conventional realities, where, as an activity, it navigates streams of consciousness vis-à -vis a relational world.

The attempt to separate conventional reality (dependent origination) from ultimate reality (emptiness), as suggested by Meyer, rests on a conflation of dependent origination with the Newtonian world. The conflation contributes to an illusion that a particular era of history, that is, the Western Enlightenment – as distinguished from Buddhist enlightenment – is *the* reality, rather than one conventional construction among others. The Enlightenment produced and rested on categories of time and space, as well as a 'religious sensibility', as discussed by Geertz, which further separated the conventional from the quantum world. In the dependent arising of the Enlightenment as a period of history, Newtonian principles of materiality, causation and mechanism were given form and grounded in a particular way of being in the world which then shaped the spatial map of the globe through practices of state-building and imperialism.

Free will and determinism are indebted to an ontology that assumes the autonomous rational agent. The concepts have no clear place within Buddhism, or Daoism and Hinduism for that matter.[2] As Wittgenstein

2 Readers interested in that debate should see Repetti (2019).

(1958) might say, the two belong to different language games: The first seeks to resolve a seeming contradiction that arises from an intellectual problem. The second accepts contradiction but seeks a practical solution to the suffering of sentient beings. The next section explores the relationship between self and consciousness in Buddhism and how the latter relates to sentience and suffering.

Consciousness and sentience

The previous snapshot highlighted the relationship between self and illusion. A further elaboration, shared by three Madhyamaka Buddhist philosophers, Nāgārjuna (2nd century), Āryadeva (3rd century) and Chandrakirti (7th century), regards the nature of the self as an activity which arises from streams of consciousness. In this conception, self cannot be reduced to streams of consciousness relating to the senses but self also has no substantial existence apart from these streams. The self cannot be grasped, yet it is the consciousness of seeing, hearing, smelling, breathing, of the sense by which life is experienced. The concept of self makes it possible to distinguish 'mine' from 'yours', to form plans, act with intention or make promises, enter into commitments or accept responsibility (Ganeri, 2012: 189). More classical conceptions of self, by contrast, reify the facts of selfhood, resting as they do on the same model as the possession of external objects.

Buddhists rely on the metaphor of fire and fuel to delineate the relationship of dependence between self or person and streams of consciousness. The relationship between fire and fuel, as discussed by Nāgārjuna (*Mulamadhyamakakarika* 10.14: 192), involves a multiplicity of possible relationships, with the fire being at one with the fuel, different from it, possessing it, the locus of it, and located in the fuel. The relationship between the self and the streams is, according to Chandrakirti, one of 'appropriating' (*upadana*), which is akin to an activation of a will. According to Ganeri (2012: 198), the word for fuel carries the sense of 'grasping, clinging, addiction', which fits with the idea of self as an activity of appropriating. The self as acting, the activity of 'I'-ing, relies on a relationship between the acting agent and the streams by which the agent is composed. The self, in collapsing into a consciousness of 'I', appropriates a space called 'mine' from which the self acts, navigates and directs the various streams of consciousness.

The self, in this conception, becomes an 'ensemble of every-changing and causally dependent processes' arising from five aggregates (material form, feeling, perception, formations and consciousness) (Gowan in Repetti, 2019: 19). There is no unchanging essence over and above these or forming a centre that holds them all together, only an appropriating self. Greed and aversion, which in the Buddhist conception are the source of suffering, arise from a dichotomy regarding what is and is not 'me and mine'.

The ability to suffer is at the heart of what it means to be a sentient being. If free will and determinism, as Zanotti (2019: 69) notes, are bound up in a struggle for freedom versus power as they relate to notions of sovereign agency, the Buddhist emphasis on the suffering of sentient beings raises a question of what constitutes suffering and how this might impact on intention and volition. Sentience is the force that is life. In classical Newtonian science, life is a chance occurrence without any inner purpose given that the organism is understood to be a complex machine, which leaves little room for consciousness. By contrast, in the Buddhist, Daoist and Vedic – as well as quantum – view, life is consciousness and consciousness manifests itself in different gradations of sentience. Further, the subjective experience of consciousness is fundamental to reproducing the world. In contrast to the mind/body Cartesian split that has characterized Western thought, consciousness is understood to permeate the body, as well as the universe, and is a necessary condition for performing any activity. In the absence of consciousness, the body would no longer function (Shanta, 2015).

A concept of 'life force' has been embraced by cultures across the world, although deemed unscientific from the perspective of classical science. The Western concept of *élan vital*, which has a much longer history, takes the form of *qi* in Chinese, *ki* in Japanese, *prana* in Hinduism and Buddhism, *gi* in Korean, *pneuma* in ancient Greece or *manitou* in the culture of the indigenous peoples of the Americas. These holistic models see humans as a part of their environment, including nature, rather than as standing outside of it. Traditions that embrace some notion of a life force also rest on a very different understanding of the mind-body relationship. The life force, or qi, is central to the holistic perspective of traditional Chinese medicine, for instance, wherein the mind and body cannot be understood as either separate in and of themselves or detached from the larger more holistic context of a person's life. The mind-body assumptions that underpin practices of health also often relate historically to traditions of warfare. Most military thinkers in ancient China, and not least Sun Tzu, devoted a few passages to the life force, which was seen as crucial to commitment and nurturing courage. While discredited within frameworks that are underpinned by classical physics, which have been dominant in the West for the last few hundred years, the life force has a place in many cultures.

The life force allows us to entertain the possibility that consciousness is present in the 'deep structure of matter' (Wendt, 2015: 111). Contrary to the conventional wisdom of classical science, Wendt (2015) argues life is constituted by an unobservable, non-material life force or élan vital, which he refers to as a 'kind of vitalism'. Life goes all the way down. Organisms are autopoietic, which means that they feed on energy to sustain their self-production in the face of thermodynamic decay (Wendt, 2015: 134).

Sentience is a product of the life force and, while taking more complex forms among humans, suffering arises, among other things, from that which threatens life. For instance, all life has the shared experience of suffering in the absence of food, which would then connect to a volition or intention to seek out food. Intention of this kind is deeply entangled in an environment, and thus impacted by the ability of the latter to actually provide sustenance, and, if it does not, the possibility of altering or changing environments. We can see this, for instance, in the changing food habits of birds. Sheldrake (2011) uses the example of blue tits in London that developed a new pattern of opening milk bottles with their beaks, a pattern which became increasingly sedimented as more and more blue tits engaged with the practice. The evolutionary change, in this case, was dependent on what happened in the past (that is, food-seeking behaviour) but involved adaptation to what is happening in the now (food-seeking in an urban environment). In so far as the blue tits were prevented from following their usual habitual path to finding food, they found a new way to reach the same goal (Sheldrake, 2011: 478). Birds would not be viewed as rational actors exercising free will, but we can see that the intent to find new sources of food relies on some form of intelligence and memory that extends to a relationship and patterns of relationship to other birds. There is no essence to the relationship given that the practice adapts to a changing context. The patterns are not meaningful with the same complexity as human practice, but one can say, as with practice of any kind, that there is nothing but the practices or the patterns themselves. Here we are reminded of Max Tegmark's (2015) statement that 'consciousness is the way that information feels when it is being processed … It's not the particles but the patterns that really matter.'

A similar logic applies to human intention, with the added complexity of language, institutions and ethics that have emerged around it. In the classificatory scheme of Buddhism, aversion to suffering becomes a source of suffering, which arises from attachment to the illusion of self and fear of harm to or loss of one's own life. I am hungry. There is a scarcity of food in my environment. My own security requires that to avoid suffering, I cause suffering for others. For example, I steal their food, or, more relevant to the context of the COVID-19 pandemic, I hoard food which means there will be less available for others.

Nāgārjuna's thief example (in Snapshot 1) is insightful in suggesting the potential differentiation of human consciousness and intention. One could engage in mindfulness, which may lead to an awareness of the impact of one's actions on another and an awareness of mutual entanglement and dependence on food. The latter might further lead to compassion arising from a recognition that self and other share a similar fate; that is, without an intake of food all life will eventually die. One can also remain in the ego state of 'This is food is mine', with little awareness of the impact of one's actions on the other. The example illustrates a spectrum of intention and consciousness that runs from the purely egoistic to the compassionate. The most extreme example of the latter is that of the Bodhisattva or enlightened one who, having overcome attachment, including to the body, acts purely with compassionate intent to the extent of sacrificing the self for the sake of others (Fierke in Koschut, 2020). Anyone may be capable of exercising free will in choosing to die for the sake of something greater than the self. However, from the perspective of an autonomous self, the preservation of individual life would be the ultimate utility to be maximized. A choice to die would thus be highly problematic.

What is the bridge that connects the life force to an experience of no-mind and, flowing from this, of action? The concept of an autonomous self, which acts more or less freely vis-à-vis an external world and exerts some causal force, poses the question of whether structure or agency is dominant. The Buddhist concepts of emptiness and dependent arising, by contrast, suggest that the two are deeply infused with each other. Agency does not consist of acting against structure but rather, in a situation of radical uncertainty which is characterized by the absence of structure, the entangled self 'knows' and is able to respond with extreme rapidity. Human knowing of this kind relies on an intelligence that is not unlike that found in other life forms in nature. As the concept is difficult to grasp, we start with a cultural depiction of no-mind and from there unpack the nature of the life force and intention as it relates to the ethical problem of killing within a nonviolent tradition. The intention here is to explore the meaning of agentless agency within a classificatory scheme provided by Buddhism, while recognizing that Buddhism is by no means a monolithic construction. Why look at a cultural and historical example that arises from Zen Buddhism, which belongs to the Mahayana tradition? Why not look at a contemporary example of Buddhist monks in Southeast Asia, who tend to belong to the Theravada tradition? Rather than embracing one or the other as more true, I defer to the claim in Snapshot 1 that every construct is a product of dependent arising. As such, none can capture truth in its fullness.[3] The example that follows concerns a practice informed by Zen Buddhism within a particular historical, cultural and spatial context, which was further shaped by the travel of Daoism from China to Japan and its engagement with Buddhism. It provides a segue into the next snapshot, which zooms in on a Daoist concept of action as actionless.

3 Those interested in the Theravada tradition should see Jerryson and Juergensmeyer (2010) and Lehr (2019).

The samurai

As stated earlier, the samurai existed as a historical phenomenon, but who and what they were as a warrior culture has been culturally produced. The samurai were originally like police who had been employed by lords, referred to as *daimyo*, to protect their land and power, but became a warrior class whose sole business was fighting.[4] One was born into the occupation and a class which became one's divine destiny. Perfection of the craft was closely tied to personal perfection, which involved both mastering a set of skills and developing superior awareness and clarity of mind (King, 1994: 113). The samurai were said to be guided by a code of honour, referred to as *bushido*, which means the way of the warrior. The ethical code was influenced by Zen Buddhism, particularly as regards the importance of meditation and the understanding of life and death. The Zen tradition (Dhyana sect) in Japan played an important role in training warriors and thus in Japan's military history. During the Muromachi period, from the fourteenth to the sixteenth century, the samurai were also an exceptionally cultured, literate and artistic elite. The Zen method of training encouraged instinctive, immediate responses by encouraging action to arise from the depths of the unconscious (Demieville in Jerryson and Juergensmeyer, 2010: 37). Zen provided a framework for entering into battle with the knowledge that one would likely die, thereby providing a way for warriors to make sense of the constant confrontation with death in a context of radical uncertainty.

The samurai code combined the Confucian ideal of the complete man as both a warrior and a scholar who not only possessed martial arts but was lettered. The balance between the military and the civic, yang and yin, resonates with the Daoist influence on Zen. Benesch (2014) argues that the ethical code of bushido was formed during the Meiji Restoration (1868–1912), during which time an idealized past was romanticized, as a response to Japan finding its way in the midst of heavy influence from the West. In this respect, what he refers to as the 'invention' of bushido was a search for Japanese national identity and a modern native ethic. The process accelerated after the annexation of the Ryukyu Kingdom (1879) and the colonization of Taiwan (1895), but particularly after the Russo-Japanese War (1904–05). The annexation of Korea (1905) signalled Japan's emergence as an empire. During this time, there was a surge of nationalistic fervour, which was reinforced by the elevation of bushido and what Benesch refers to as the bushido boom, during which bushido was redefined for militaristic and propaganda purposes and promoted a pro-expansionist samurai ideology that

[4] They were known for their courage, endurance and loyalty, but were remembered primarily for their prowess in battle.

centred on loyalty to the emperor and the imperial nation. The Japanese martial spirit was understood to be more important than material capabilities or technology. The construction, which evolved over time in relation to several foreign encounters, was an attempt to push back at foreign influences, and not least the extreme individualism of the West, by making bushido the core of what is uniquely Japanese.

No-mind and killing

The Last Samurai (Zwick, 2003) is a film that was inspired by the life of Saigo Takamori (Ravina, 2004). It should be mentioned that Katsumoto, who represents Saigo Takamori in the film, was somewhat idealized. Takamori advocated the military invasion of Korea in 1873, in the midst of growing dissatisfaction among the samurai class.[5] The focus of the film is the confrontation between Western and samurai culture and how the samurai became victims of the advanced weapons technology of the West. It says little about the mood of Japanese imperialism towards neighbouring territories that emerged from this confrontation. The cultural production romanticizes a battle between 'good' guys and 'bad' guys and, in the process, neglects some of the more imperialist tendencies of the samurai themselves, which would then be magnified as samurai ideology mixed with Western technologies. Nonetheless, the stark contrast in the film, while exaggerated, is useful for highlighting a number of points.

The film begins with an American captain with a highly successful track record in battle against Native Americans in the US in the nineteenth century. Despite his success, the captain is also severely traumatized, as reflected in his heavy drinking. Because of his successful track record in killing Native Americans, he is invited by the Japanese emperor to train a group of Japanese conscripts in the strategies and tactics of Western warfare to the end of suppressing the samurai. Samurai means 'service', and in this case, paradoxically, the leader of this ancient tribe, Katsumoto, understood his service to be to the emperor. Following a military engagement between the still untrained Japanese soldiers and the samurai, which results in the defeat of the former, Katsumoto is so impressed with the captain's fighting skills that he is taken prisoner rather than beheaded. Katsumoto's intention is to learn about his new enemy. The captain is nursed back to health by the wife of a man he had killed, during which time, denied of alcohol, he has to confront flashbacks and haunting images of having

5 As Dudden (2006: 49) notes, 'The famous revolutionary Saigo Takamori declared Korea's rebuff a greater affront than the unequal treaties with the West, and he urged war with Korea – a war that would enable him in midlife to again wield arms.'

killed Native American women and children during his earlier raids in North America. During the process of healing, he begins to observe the samurai's approach to fighting, and its relationship to a particular way of life, in which a warrior devotes himself entirely to a set of moral principles, bushido, and to calming the mind. At a certain point the captain attempts to learn this method of warfare. As a proven warrior in another context, he is determined to succeed but continuously finds himself thrown to the ground. After being battered for the umpteenth time, a young boy said to the captain, 'Too many minds', to which he elaborates, 'Mind the sword, mind people watching, mind enemy'. The boy's conclusion is 'no mind', at which point the captain, as he begins to grasp the point, adopts a totally different stance toward his opponent.

In this scene the Buddhist concept of no-mind intersects with the Daoist concept of wuwei or actionless action. The latter will be discussed in more depth in the next snapshot. At issue here is how action flows from no-mind. The scene in the film provides insight into the transition from 'many minds' to no-mind. The boy's reference to many minds suggested that the captain's previous failure arose from a burden of overly rationalized mental multitasking. At any one moment he simultaneously thought about how best to use his stick and what his opponent was going to do, not to mention how the audience was responding, all of which took too much time given the speed of engagement and the distraction from the task at hand. The concentration and consciousness of the captain was fractured.

Self-consciousness constructs a division between mind and action, separating the self from its environment and thereby creating a bifurcated reality (Yamamoto, 2000: xx). Diverse narratives related to, for instance, the direction of any forthcoming blow, fears of being harmed or the public humiliation of defeat add further layers of thought. In the multiple possible narratives and the movement of mind through them, there is a further separation between self-action and the objective, that is, to defeat the opponent, who in a 'many minds' framework is a separate other. This manner of thinking is too cumbersome to be effective in a context of radical uncertainty, where death can come at any moment. Rational thought depends on the premises embedded within specific narratives. Particularly in a context that is unfolding at great speed, the agent is faced with the difficulty of juggling what is ultimately a fragmentation of self that arises from the multiplicity of minds. The juggling act comes at the expense of attentiveness to the sounds and other stimuli provided by the opponent and the larger environment. The attempt to impose form on the mind for purposes of control ultimately results in a loss of control.

What is powerfully portrayed on the screen is the moment when the captain's mind empties, as he adopts a different relationship to his mind and environment and thus to what his opponent is doing. The reorientation is followed by a series of very rapid actions to suppress the other before he can do harm. In this shift to no-mind, we see the paradoxical combination of, on the one hand, a contemplative aspect, which is acquired through practices of meditation, and purposive action in a context of fighting. The mind is turned into a reflective mirror that is able to separate thoughts and feelings from aspects of the immediate environment or the opponent's state of mind. If the mind-mirror is clouded by one's own concerns, hopes, fears or perceptions, it is impossible to clearly sense what is happening around you (King, 1994: 113). Calming the mind is necessary for beginning to act freely and spontaneously, to attack without the need for either deliberate thought or paralyzing fear. The former relies on consciousness that permeates not only the mind but the body as well, and a mind that is emptied of self and thus of self-consciousness. Chuang Tzu (Fung, 1970) referred to the quiet centre that does not change in the midst of constant turbulence or activity as 'tranquility-in-disturbance'. Action from a place of emptiness is an ability of the genuine human being who is one with the Dao, as discussed in the classic writings of the Daoist Zhuangzi (Ziporyn, 2009: 40–5).

A philosophy of mind is central to successful swordsmanship. The samurai had to be ready for instantaneous action and possible death at any moment. They developed an uncanny ability to detect danger from the slightest bodily or facial indications of the opponent. Truth wasn't sought in scripture or formal learning but through the development of the self and its capacities. It was thus existential rather than intellectual, and built on the Daoist contrast between yin and yang, contrasting elements in tension with each other. The tension was creative rather than destructive, and through practices of meditation, a stable platform of the mind was found in the midst of whirling motion. Meditation, combined with advanced training in swordsmanship, made spontaneous, visceral, instinctive and exceptionally rapid action possible, such that a warrior could take on a threat even when outnumbered. Fundamental to the practice was a concept of no-mind and a place of perfect emptiness from which action arises, as reflected in the poem (King, 1994: 174):

> Victory is for the one,
> Even before the combat,
> What has no thought of himself,
> Abiding in the no-mindness of great origin

The readiness to die and the strength and intensity of the commitment must be constantly reinforced through meditation, which was the focal point of bushido existence, in which death becomes essential to the meaning of life (King, 1994: 177). Death in this meaning relates first and foremost to death of the ego, or the Zen concept of *muga*, rather than necessarily physical death. There is a saying from *Zenrin-kushu* ('Anthology of Passages from the Forests of Zen'), a collection of writings used in the Rinzai school of Zen, that to preserve your life you must kill it. Kill it off completely, and you will be at peace for the first time (Yamamoto, 2000: ix).

Intuition as useful entangled knowledge

The possibility of action from a place of emptiness highlights a number of issues, not least the paradox of action without rational thinking and action that is ultimately *more* effective at prevailing in the face of danger. Reason, as a logical linear process that links thought, action and objective, may not ultimately be rational if the internal and linear process detaches the self from knowledge of the larger entangled whole; and if indeed there is a more direct and speedy manner in which to engage. From this perspective, what has been considered rational is ultimately irrational, in so far as it stands in the way of achieving one's objectives through more direct knowledge, that is, intuition. The film effectively depicts a relationship between mind and action that is central to the martial arts more broadly, and which at first glance would appear to express a mystical experience.

Taking intuition seriously requires stepping out of a world in which parts are understood as discrete material objects, lacking in connection or consciousness. It further requires rethinking the relationship between matter and consciousness to go beyond the mutually exclusive terms of realism and idealism in which the body, or the perceptions generated by the individual mind, are understood to be separate and grounded in a universe within which, on the one hand, consciousness is merely epiphenomenal to matter (realism) or on the other, objects are mind dependent (idealism). Consciousness and matter, or wave and particle, are always in a dynamic relationship to one another. As the French philosopher Henri Bergson (1911: 116) states:

> We maintain, as against materialism, that perception overflows infinitely the cerebral state; but we have endeavoured to establish, as against idealism, that matter goes in every direction beyond our representation of it, a representation which the mind has gathered out of it, so to speak, by an intelligent choice. Of these two opposite doctrines, the one attributes to the body and the other to the intellect a true power of creation, the first insisting that our brain begets representation and the second that our understanding designs the plan of nature.

Bergson makes a distinction between a notion of reality that appears as immediate intuition and one that arises from an adaptation of the real to 'useful knowledge' that shapes the practices of social life. Pure intuition, in his argument is an 'undivided continuity' (Bergson, 1911: 117), the entangled whole of the universe. Like Buddhist emptiness, Bergson moves away from the either/or, idealism/realism dichotomy as part of a critique of empiricism, which emphasizes the importance of evidence that can be verified through observation. Empiricism is underpinned by a notion of phenomena as separate and discontinuous in time. However, what is posited as real knowledge is in fact, according to Bergson, 'useful knowledge' of our own construction, and thus contingent. By contrast, intuition, in its original purity, provides the potential to recover contact with a reality that is tied to 'human experience' at its source (Bergson, 1911: 118). The two cannot ultimately be separated. The world within which we move, while arising from emptiness, takes form through the various practices by which we engage with it, which includes the continuous making of distinctions. Intuition and consciousness belong to the whole, and that which arises from it cannot be separated out entirely without losing the very essence of life, that is, the élan vital, or in Japanese, ki.

The intention to defeat an enemy belongs to a tradition and the customs of warfare, which require training and the development of the skills needed by the warrior. Skilful action of any kind, whether *aikido*, dancing or playing the piano, is in part a matter of technique, but can never be purely a matter of technique. It is also the loss of ego in the performance and the expression of the spirit of the practice that distinguishes the merely competent from the sublime (Knightly, 2013). From this angle, insight and intention, on the one hand, and on the other, spontaneous action, can be distinguished from the explicit choice arising from free will. Insight and intention arise from mastery of the conventional combined with surrender to something that is larger than the self. The perfect embodiment of practice, that is, the ability to dance or fight effortlessly precisely because one has mastered a skill, comes together with intuition, an attunement that is perfected in practices of meditation and mindfulness. For the warrior, the practice of meditation is itself a skill that makes it possible to empty the mind (no-mind) in the rapidly changing context of battle. No-mind requires simultaneously detaching from the conventional while engaging with it. The infusion of the life force or ki into the body is fundamental to the effortless movement. The skilled Zen warrior, who has developed skill to perfection, acts within a particular order of war and, at the moment of action, is attentive to the environment in a way that can only be attained through emptying the mind of ego, no-mind. On the one hand, a good warrior is able to melt into the landscape, becoming a part of it (Deshimaru, 1982: 39). On the other hand, the perfection of skill involves more than technical expertise. It is further the 'perfection of a unique and ultimate skill: the skill of becoming a fully realised human being and embodying the Way in the full range of one's actions' (Slingerland, 2000: 313).[6]

The ethics of violent intent

The last section hinted at the relationship between no-mind and acting with intent, but the latter requires further specification, particularly as it relates to the ethical import within Buddhism of acting with *violent* intent toward a sentient being. I have elsewhere explored this question in relation to harm to the self in the form self-immolation by fire in the contexts of Vietnam (Fierke, 2012) and Tibet (in Koschut, 2020). Both suicide and killing are prohibited in Buddhism. There are, however, exceptional circumstances in which either act, if undertaken with right intention, and by a bodhisattva or enlightened one, is understood as an offering and sacrifice to the Buddha that transcends moral precepts.[7] In this respect, the prohibition on harming life is adaptable, depending on the purity of intention. The central point is that the act should neither be that of a desperate individual nor should it emerge from the aspiration for one's personal enlightenment, perhaps in another life, both of which represent forms of self-grasping or attachment (Garfield, 2001: 513). The act must arise from an intention to alleviate the suffering of sentient beings, that is, a compassionate intention. If the intention is contaminated by the poisons of desire or aversion, they are, by definition, unwholesome (Kovan, 2014: 781).

There is space within Buddhism, and Mahayana Buddhism in particular, of which Zen is a branch, for exceptional acts of killing. In some respects, while no less paradoxical, a compassionate intention to kill the physical self is easier to comprehend than one that involves killing others, except perhaps in the case of extreme suffering that would be relieved through death. In Western culture, animals are often killed to save them from suffering, but an act of this kind, for instance in the form of euthanasia, remains controversial in most countries. The most acceptable acts of killing are symbolic rather than physical. In the earlier discussion of the samurai, it is the ego self and attachment to the body that in the first instance dies. The latter does not by definition mean physical death, although an absence of ego may assist in coming to terms with this possibility.

The compassionate intention belongs to consciousness of one's entanglement with others, which is larger than the impermanent, egoistic self. The intention is the liberation of all sentient beings from suffering and

6 The Way refers to the Dao, which is discussed in the next snapshot.

7 In the Mahayana tradition, action is secondary to intention. The deed will be considered pure if the intention was pure, but if the intention was impure the deed will also be (Kleine, 2006: 162–3).

samsāra (the never-ending cycle of death and rebirth). The actor connects to a transpersonal notion of self rather than self-interest. As the killing becomes physical, ethical questions arise. Self-immolation by fire harms the physical self but not others.[8] The objective of compassion is not unlike the symbolism of Christ's death, although in the bodhisattva's practice of self-immolation, death is self-inflicted rather than inflicted by authorities. The two practices share a family resemblance while arising from distinct dependent originations. Self-immolation is a performance of compassion through which the bodhisattva is produced.[9]

That an enlightened one would kill others is far more difficult to reconcile with a tradition rooted in nonviolence. Madhyamaka thinkers, such as Bhaviveka, Candiakirti and Santideva, make the basic point that bodhisattvas can undertake acts, including killing, that would ordinarily be forbidden. These acts will have merit so long as they remain compassionate (Jenkins in Jerryson and Juergensmeyer, 2010: 68), and indeed, compassionate violence is accepted and common to ethics across the Mahayana traditions. 'Do not kill' is one of the grave bodhisattva's precepts in Zen Buddhism, yet it can be overridden for compassionate reasons. The intention to act from ego or compassion is what is most crucial in determining merit. In the larger historical and philosophical narrative, the bodhisattva is a transpersonal aspiration which is at odds with the idea of an act motivated by a personal search for enlightenment (Garfield, 2001: 517–8) or any notion of intention that relates purely to individual interest.

Killing and just war

The rule of nonviolence in Buddhism is adaptable, and indeed there are sufficient historical examples of this bending, whether in relation to self-immolation or armed resistance or warfare. What is crucial is the intention of the action. Intention is less about the reasons that an individual might give for his or her action than the context of the act and how this relates to a self-interested or compassionate motivation. The thoughts that flow through the mind of any one actor or observer may be multiple and fleeting, but the question of whether an act is one of self-grasping, compassion or defence is of a different kind. The reason for an action or its justification can thus be distinguished from the intention it expresses. Intentionality is woven into the narrative surrounding the act, which connects the various thoughts that

8 Although harm of a different kind may be experienced by others, for instance, relating to loss of a loved one.

9 In Mahayana Buddhism, *bodhisattva* refers to anyone who has generated *bodhicitta*, a spontaneous wish and compassionate mind to attain Buddhahood for the benefit of all sentient beings.

flow through any one individual mind with those expressed more publicly in socio-political discourse (Fierke in Koschut, 2020). Yet, the skilled action arises from a mind that has been calmed with the cessation of concepts.

The attempt to judge the status of any one performer as bodhisattva seems contrary to the philosophical underpinning of dependent origination and emptiness. Dependent origination, and the absence of intrinsic identity, point towards an understanding of intention that is similar to that expressed by Wittgenstein, according to which bodily action becomes an expression of intention, of the customs, institutions and practices by which it has meaning (Wittgenstein, 1958: para 337). In this respect, the skills of the warrior, relating both to meditation and fighting, arise from the customs and institutions of Zen Buddhism and the causes and conditions by which a particular war has come to be justified, including the customs and institutions of a specific context. Fighting in defence of a population can be considered selfless.

Various schools within Buddhism have justified acts or habits that are contrary to the Buddhist precept of non-killing, which underlies its code of ethics (Demieville in Jerryson and Juergensmeyer, 2010: 33). My purpose is not to draw conclusions about what is ethical, which is heavily dependent on context. It is rather to explore what the cultural representation of no-mind says about the ethical significance of killing in war. The paradox is that we can't distinguish any one ethical articulation from the dependent origination of concepts within a particular time and place. The most pressing question in Buddhism, with its nonviolent underpinnings, is how violence of any kind can be justified in light of the concern for the suffering of sentient beings. The final section has the more limited objective of exploring the relationship between the ethics of killing as it relates to particular forms of warfare.

American no-mind and healing

This snapshot ends with a further example that approaches no-mind and the ethics of warfare from a somewhat different angle. The context regards the adoption of techniques related to no-mind in the healing of Western soldiers who suffer from PTSD as a result of fighting in Afghanistan and Iraq. Within the US military, complementary and alternative medicine (CAM),[10] which includes mindfulness, among other things, has increasingly been tested for its effectiveness in trauma management. CAM is compatible with the US Army's development of resilience training, to the end of preparing soldiers to more effectively deal with the stresses of war, and relies heavily on behavioural and cognitive therapies and 'positive psychology' as well. Examples of the former include mindfulness-based mind fitness training,

[10] Also referred to as complementary and integrative medicine.

which is widely available to soldiers in the US and has the mission 'to protect soldiers as they prepare for deployment and experience stressors',[11] and to improve mission performance and reduce rates of PTSD, depression and anxiety in soldiers upon return (Meredith et al, 2011: 141). Resilience, like any other concept, has been defined in diverse ways. Meredith et al (2011) have identified three types of definition that alternatively describe resilience as 'a process or capacity that develops over time' (basic), the ability to 'bounce back', adapt or return to a baseline that preceded the experience of adversity or trauma, and the ability to grow following an experience of adversity or trauma. Resilience requires positive coping, affect and thinking, as well as physical fitness, realism and an ability to control one's own behaviour. It also rests on a positive potential for connectedness, emotional ties and feelings of belonging, whether to the family, military unit or community.

The concept of resilience has been challenged by more critical scholars of international security, with arguments that resonate with the mindfulness debate (for more see Snapshot 1). Howell (2012), for instance, notes that resilience programmes, which teach soldiers 'positive psychology' and ability to 'stop catastrophizing', place responsibility for developing (or not developing) PTSD with the soldier, thus shifting the onus away from military institutions to the individual. If one develops PTSD, it is a sign that they are not sufficiently resilient and unable to 'think positively' during catastrophic events. She further argues that the resilience model is socially and economically useful given the tremendous financial costs of disability and compensation claims in relation to Afghanistan and Iraq. In the US, in the absence of a comprehensive civilian national health service, the costs of benefits and long-term care for veterans over the next forty years has been estimated at 600 billion to 1 trillion dollars (Bilmes, 2021). Resilience training primarily emphasizes prevention.

While the traditional scientific literature seeks to establish whether these methods are effective, arguments within critical security studies raise questions about the self-interested reasons for the shift to an emphasis on resilience. Both types of argument highlight the application of these new technologies within Western military institutions rather than the cultural underpinnings of the practices. As Baer (2003) notes, Western researchers and clinicians who have experimented or applied mindfulness practices in the context of mental health treatment programmes usually divorce these skills from the religious and cultural traditions from which they originate. The intention is to return soldiers to the battlefield. As argued in the mindfulness

[11] The Warrior Resilience and Thriving Program, located at Fort Sill in Lawton, Oklahoma, is somewhat more explicit about its links to Eastern traditions, drawing, among others, on Japanese Bushido and other warrior cultures (Meredith et al, 2011: 154).

debate, the practice contributes to the reproduction of a particular kind of war that is motivated, among other things, by the desire for profit.

As such, we return to the subject with which this section of snapshots began, that is, the introduction of mindfulness in a context, such as the US, which is a very different culture than those from which practices of meditation and mindfulness, or any of the healing forms mentioned at the start of this snapshot, originated. Medicine and warfare have historically developed hand in hand (Howell, 2014). Healing and fighting practices arose in the US and Asia from causes and conditions that belong to very different cultural contexts. Nonetheless, in the aftermath of Iraq and Afghanistan, the ancient healing traditions produced 'small miracles' in the treatment of PTSD (Miles, 2008). The scene from *The Last Samurai* demonstrated the use of no-mind in a context of fighting of an earlier kind. The contemporary use of mindfulness in the military is for purposes of healing wounds suffered by soldiers in a very different context of warfare, and thus gives rise to both contradictions and potentials. The film, once again, provides a useful foil because it intertwines two distinct cultures of warfare and the consequences of their engagement with each other at an earlier point in history.

PTSD also has a role in the story. Several insights from the storyline provide ethical food for thought. First, the American captain, like many soldiers who fought in Vietnam, Iraq or Afghanistan, suffers from PTSD, although the category, as a product of the post-Vietnam era, did not exist in the nineteenth century setting of the film. The captain's suffering, and the alcoholism attached to it, arose from flashbacks of his participation in the indiscriminate killing of Native American women and children. The traumatic images or narratives that cloud the mind of contemporary soldiers may be of many kinds. They may arise from flashbacks of bombardment and/or death, as well as experience as a perpetrator or victim. As for the nineteenth century captain, PTSD may be a response to violations of the soldier's own moral order and betrayal of what is right (Shay, 2003: 6, 15), arising, for instance, from participation in indiscriminate violence against innocent populations. The experience of 'moral injury', arising from life and death decisions in highly uncertain and rapidly changing conditions, may be a further factor. No-mind, in the context of PTSD, is about emptying the mind of traumatic images and accompanying negative narratives, often of humiliation or betrayal by the self or others, so that they don't immobilize the soldier, thereby allowing him or her to take back control over the body. Trauma represents a fragmentation and loss of radical relationality (Edkins, 2006: 102), which shatters everyday feelings of safety within the conventional world of experience. By contrast, the no-mind of meditation provides a different way to experience a space of emptiness beyond words, which points toward a possible restoration of relationality. But there is a potential paradox, or contradiction, in the contemporary context: that which is intended as preparation for a return to the battlefield may instead set the soldier on a course that raises fundamental ethical questions, if not about warfare itself, then about warfare in highly industrialized conflicts such as Afghanistan and Iraq.

PTSD, coined in the context of the Vietnam War, has been a significant issue for Western militaries, particularly in the US. Soldiers who fought in Vietnam, Iraq or Afghanistan were not directly defending kith and kin. In the case of Vietnam, they were not welcomed upon their return, given questions about whether the war they fought, and the means of fighting it, were morally right. Moral questions have also been raised in relation to Iraq, where the legality and thus legitimacy of the invasion and subsequent occupation were called into question, not to mention the disproportionate number of dead Iraqi civilians. The US Army has been experimenting with forms of mindfulness and other techniques such as yoga, tai chi and reiki in treating the PTSD of soldiers who fought in Afghanistan and Iraq. Paradoxically, Western soldiers are now using methods derived from cultures which, in an earlier time, were defeated by highly industrialized forms of warfare driven by the desire for profit. While the concept of trauma was originally associated with physical injury in war, PTSD is a byproduct of the psychology of the individual mind, which emerged in the context of industrialization and World War I, and innovations such as high-speed rail and total warfare (Fierke, 2015).

Mindfulness helps contemporary soldiers find some peace from the flurry of images and narratives that now imprison them and, like the memories of the nineteenth-century American captain, place an unbearable burden on their daily existence. On the one hand, the captain violated his own moral order and was himself transformed by the guilt of having massacred innocent women and children. He thus suffered with his victims, albeit in a different way: having harmed, he was himself harmed in so far as he was psychologically debilitated by his actions. His primary relationships were to the dead and to a bottle. On the other hand, as he began to heal and develop a different approach to mind, he changed. The conventional engagement with 'many minds' was with time replaced, through the practice of mediation, by no-mind.

From the place of emptiness, the self is no longer engaged in a battle of self-protection against the onslaught of perceived dangers, or pre-occupied with combatting these dangers. Once the mind is emptied and calmed, one becomes more aware of the surrounding environment, the sounds, the movements and the sights, and thus more able to respond in a way that is attuned to that environment, including with greater compassion towards others. As the internal battle is stilled, the warrior is also better positioned to respond defensively with speed to dangers that may arise. Ancient Eastern approaches to warfare, from that of Sun Tzu to the samurai and the martial arts, begin from this point of emptiness. When the mind is emptied of its own pre-conceptions, and no longer threatened by negative images or narratives of shame, betrayal, guilt or fear, it opens to the evidence and stimuli of the surrounding environment. Because more attuned, the agent has greater ability to respond directly and more effectively to actual threats, or, in the absence of threats, to avoid putting the other on the defensive unnecessarily.

The Last Samurai depicts a confrontation between two traditions of warfare. One involved disproportionate force against Native Americans and, in another part of the world, the samurai in the name of Western capitalism and imperialism. While recognized as extremely powerful warriors, the samurai were unable to compete with Western heavy artillery and a strategy of overwhelming force. The decision of the Japanese government to acquire Western technology, at the expense of their own samurai traditions, was framed in terms of needing to work with the Western powers in order to maintain their sovereignty, while, paradoxically, giving up an element of their culture in the process. The conflict of Japanese cultural identity contained within this confrontation is personified in the person of the emperor, who mourns the loss of the samurai as a representation of Japanese culture while he, with some reluctance, concedes to the demands of the Europeans and Americans. The film depicts the clash between the skilled warfare of a tribal culture, influenced by Sun Tzu (who will be the focus of Snapshot 4), and the disproportionate firepower of the new Japanese infantry.

More recently we can point to the aggression of the Japanese military, and not least the kamikaze pilots during World War II, influenced by the bushido culture, as well as the large payloads of bombs dropped on Japanese cities by Allied forces, including atomic weapons, and the extensive bombing campaigns later in Vietnam, and more recently still in Afghanistan and Iraq. The paradox of no-mind looks different in this highly industrialized context of war. For the samurai, no-mind was part of the intelligence with which the warrior fought effectively, even when outnumbered, in a highly uncertain and rapidly developing context in response to attack. Mindfulness as a technique for treating PTSD, in the context of the modern American military, contains an element of no-mind which allows the soldier to respond with greater resilience to an environment of warfare. The warfare revolves around an asymmetry, which in the contemporary contexts was defined by the relationship between a superpower and 'terrorists'. The asymmetry is marked by highly destructive weapons, particularly those of the US.

What does this somewhat circular narrative say about how practices of emptying the mind relate to both the prevention or management of trauma and its potential implications for the ethics of war? For the captain, the practices of Western warfare became a source of trauma. Killing for purposes of ego or self-interested gain (he was a mercenary), to remove a people (Native Americans), defining them as an other who must be eliminated, became a tremendous burden on the soul of this character. When the captain spoke of his flashbacks to Katsumoto, claiming they were something all soldiers experience, Katsumoto replied, 'Only if they are ashamed of what they have done.'

The samurai approach to war involved a form of fighting that was attached to a moral code of honour, including some practices that would be viewed as abhorrent in the West, not least the beheading of the defeated and suicide in the face of one's own shame of defeat. The fighting itself, however, engaged the warrior in a more direct relationship with his environment, and arose from an explicitly defensive stance. However, while recognized as extremely powerful warriors, the samurai were unable to compete with Western heavy artillery and a strategy of overwhelming force. The warrior culture thus fell victim to Western advances in weapons technology.

As the captain developed no-mind and learned to fight in a new way, his PTSD subsided. But he also was changed. Still a warrior, he went into battle at the side of Katsumoto, fighting against his previous colleagues in the US Military. The question is whether the adoption of these Eastern techniques, including mindfulness, potentially opens a door of a similar kind for contemporary American soldiers. Will it provide the impetus for some to look further into the philosophical traditions of Buddhism or Daoism for different kinds of answers to their suffering? Like in the thief example from the first snapshot, the practice in itself has the potential to give rise to greater reflection about the relational nature of being, and thus to ethical reflection on the nature of war.

Conclusion

Snapshot 2 zoomed in on the relationship between sentience and consciousness, as well as the suffering of life and the ethical paradox of a nonviolent tradition being held by practitioners of violent warfare. Distance was taken from Western debates about free will and determinism in order to explore a Buddhist concept of no-mind on its own terms. Drawing on a contemporary depiction of an earlier encounter between Eastern and Western cultures of warfare, the meaning of Zen Buddhist no-mind as it relates to fighting was explored, as well as the potential significance of no-mind for healing the PTSD of contemporary American soldiers. The confrontation between two cultural depictions of mind in the film *The Last Samurai* raised ethical questions about the nature of war. One might conclude that a Buddhist ethics of war would start with the principle of no war. Buddhism is widely associated with nonviolence and compassion. But there are also historical and contemporary examples of Buddhist warfare, not least the samurai culture of Japan or of contemporary monks in Southeast Asia. The paradox remains a paradox.

The ethical paradox is expressed in a puzzle from the Zen Buddhist Dogen. The puzzle revolves around the need to choose between culling a population of deer that have grown too large or letting them live, knowing that they would harm the ecology of a small Scottish island which was being devastated by the overpopulation of deer. What is the ethical answer to the need for action in a context where loss of life will be the outcome of any decision? James (2004) argues that the Zen Buddhist could be justified in choosing to cull the deer if she did so out of compassion for the deer or out of concern to preserve the island's ecosystem. Balance is the key issue here. There are plenty of examples of death and harm in nature and, in this respect, death, the culmination of thermodynamic decay, is a part of life. There is, however, a difference between death as a part of the natural balance that restores itself and death from violence for instrumental ends and profit.

Modern ethics in the West has been primarily concerned with providing rules or first principles that set out how we ought to act. Acts can then be measured against the rule to determine the degree of correspondence. Such practices will not be effective in a context of radical uncertainty. If the defining feature of modern ethics has been a concern with what is right, Zen Buddhism, according to James (2004), represents a form of virtue ethics, which has a counterpart among the ancient Greeks. Virtue ethics is concerned less with good outcomes than with what constitutes a good life, or in the case of Buddhism, an enlightened life. An enlightened life might include qualities of acting mindfully, with compassion and integrity. Paradoxically, the self that seeks the good life is empty of essence, impermanent and, in the context of this snapshot, engaged in violence and war. A Buddhist virtue ethics provides practical wisdom about acting differently that flows from seeing the world differently, free of distorting attachments. Detaching from one's separateness and assumed intrinsic identity comes with a potential to be more selfless and compassionate. The path can't be encapsulated in a set of rules but is instead developed through skill and practice, and not least through calming the mind and developing the power of concentration.

SECTION II

Complementarity and Yinyang

SNAPSHOT 3

Action/No Action

> Do that which consists in taking no action; pursue that which is not meddlesome; savour that which has no flavour.
>
> *Daodejing*, Chapter 63

The COVID-19 pandemic gave rise to stark contrasts and reversals. The mass encounter with impermanence and loss of life went hand in hand with an appreciation of life and the importance of loved ones, now separated by lockdown restrictions, even at the moment of death. Images of overflowing hospitals and mortuaries were juxtaposed with protestors demanding the freedom to ignore restrictions necessary to defeat the virus, including the wearing of facemasks. With the US elections in 2020, the outgoing president spoke of having made America great again while claiming that the election had been stolen; the incoming president spoke of grief and the importance of compassion and the need for unity, not only at home but abroad. The new year saw a largely white and male mob, incited by a Trump rally, storm the US Capitol Building, threatening senators and representatives, desecrating its premises and leaving six dead, while threatening a new civil war. The small number of police contrasted starkly with their strong presence during the earlier Black Lives Matter protests in Washington DC, once again highlighting racial inequality in America. The stark contrasts were also evident in the environment. On the one hand, fires raged in California and Australia and hurricanes in the US, Caribbean and Pacific, among others. Populations were faced with the task of both protecting or escaping their homes and fending off disease. On the other hand, the global lockdown due to the pandemic brought an emptiness and calm to the experience of nature, as well as a greater ability to breathe.

The previous snapshot explored the inseparability of emptiness of mind and action for the Buddhist warrior, a concept that would appear to be a contradiction in terms. A similar problem is at the heart of this snapshot, but we reposition the apparatus slightly to zoom out to the larger environment, both as context and natural phenomenon, to gain some understanding of the stark contrasts and reversals that characterized the pandemic. The analysis that follows brings the quantum concept of complementarity to the Daoist *yin* and *yang*, asking what the latter has to say about processes of differentiation at the macroscopic level, the place of humans within nature and what it means to act from a place of paradox and contradiction.

The metaphor of navigating a boat on extremely rough waters provides an entry to part of the problem as posed here. The image contains three parts: nature, as represented by the water, humans who are trying to navigate the water, and the boat as the apparatus for doing so. One question is how the relationship between nature and the human artefact is understood, including the status of the human vis-à-vis both. While the water is the most visible aspect of nature, its movement is propelled by the swirling air. The vessel may be impacted by other less visible phenomena, fauna and fish, or even whales or hidden land or icebergs in its midst. Artefacts from earlier shipwrecks, and the human remnants left behind, may also be lurking beneath. The larger context of nature raises a question about the relationship between the visible and the more hidden dynamics of the rapidly undulating surface waters. A further question is how to act in conditions where the waves and extreme winds are jerking the boat back and forth. As it tips too far in one direction, momentum reverses to the opposite side. The captain has no control over the larger environment through which she navigates, but she also is not passive. She works with the oscillating movements to keep the boat in some kind of balance so that it does not capsize, while steering it toward a destination. She is a part of the boat as she is part of nature.

The purpose of this snapshot is to reposition the apparatus (the parallel between quantum physics and Daoism) to explore these dynamics. Within Daoism, the polar opposites yin and yang are the dynamic force of the universe; the concept of complementarity, articulated by the physicist Niels Bohr,[1] refers to the dynamic relationship between particle and wave. In designing his family coat of arms, Bohr incorporated the ancient Chinese symbol of the Dao.[2] The inscription reads *contraria sunt complementa* ('opposites are complements'). Opposites are complements rather than logical contradictions, as they are often understood in European philosophy (Wang, 2012: 8). The physicist Leon Rosenfeld (1963: 47) asked the Japanese

1 The Copenhagen interpretation is considered the 'standard' interpretation, and is widely accepted by physicists.

2 In 1947, Bohr, the father of the Copenhagen interpretation of quantum mechanics, received the Order of the Elephant from the Danish government. It was customary for those receiving the order to carve their family coat of arms into a wall of fame. The Bohr family did not have a coat of arms so he designed one himself (Harrison, 2000-2002).

physicist Hideki Yukawa whether Japanese physicists had as much difficulty grasping Bohr's complementarity as their Western counterparts. Yukawa responded: 'No, Bohr's argumentation has always appeared quite evident to us … you see we in Japan have not been corrupted by Aristotle.' Yukawa here refers to Aristotle's law of non-contradiction, which states that opposing principles cannot both be true at the same time.[3]

Bohr's complementarity asserts that the same object of knowledge may have complementary properties; however, knowing one will exclude knowledge of the others. The result, as the physicist Heinz R. Pagels (1982: 94) notes, is that 'we may therefore describe an object like an electron in ways which are mutually exclusive, e.g. as wave or particle – without logical contradiction.' The contradiction or paradox is central to quantum phenomena. The Austrian-Irish physicist Erwin Schrodinger's famous cat experiment was an attempt to illuminate the apparent logical absurdity of quantum mechanics, given that mutually contradictory states such as death and life could coexist. Quantum computing allows for outcomes that express yes and no simultaneously. The parallel between quantum complementarity and yinyang was introduced in Snapshot 2. The key question here regards the significance of contradiction for understanding action. The Daoist concept of actionless action or *wuwei* only makes sense when situated within the contradictory dynamics of yin and yang. The central problem is one of how polar opposites generate not only contradiction but dynamic potential, and how one seizes the potential to act with the tide, rather than against it, thus achieving the maximum effect with minimum effort.

The following develops the Daoist concept of actionless action within a discussion of the global environment. As a snapshot of the 'same' phenomenon from a different angle, the argument is distinguished from that of the last section in several respects. First, from no-mind as it relates to a concept of no-self, emphasis shifts to a conception of action, or more specifically of actionless action (wuwei), which more firmly places the agent in a context and expands its multidimensionality. In doing so, I take a further step away from the language of agency, given its association with the autonomous individual, and instead refer to 'actants'.[4] Second, while Snapshot 2 explored a question of how the mind might engage effectively

[3] The discussion of the law of non-contradiction can be found in Aristotle's *Metaphysics IV (Gamma)*, Three to Six and especially Four. This is one of his firmest principles in so far as the law is seen as a requirement for distinguishing things and essences. There are three versions of the law of non-contradiction in Aristotle. The first refers to things that exist in the world (ontology), the second to what we believe (doxastic) and the third to assertion and truth (semantic).

[4] As will later be discussed, this term belongs more to the post-humanist literature, and to Latour (2004) in particular, but, like the concept of assemblage (Deleuze and Guartarri, 1987) I put it to use with specific reference to Daoist thought.

with a rapidly unfolding context, the concern here is to highlight how the environment, as a context of action, shapes the potentials available to actants, both human and non-human, without determining them.

Yinyang illustrates the emptiness/form relationship, as well as the dynamic process of differentiation that emerges from it, which is a source of creativity. The frequent association of Daoism or actionless action with no-action or passivity, I argue, arises from a failure to take contradiction, or the dynamic interplay of polar opposites, that is, yinyang, into account. The first part of the snapshot unpacks the relationship between environment and nature within a Daoist framework. The second part explores the importance of yinyang as, on the one hand, polar opposition and contradiction, and on the other hand, as presence and non-presence and as a dynamic force of differentiation. The backdrop helps to clarify the meaning, in part three, of wuwei as far more than 'no-action' or passivity. The final part returns to what this potential might mean in a context of pandemic and environmental change.

Environment and nature

While environment and nature may seem to be two straightforward and interchangeable concepts, the relationship changes within different structures of meaning. Cheng Chun-Ying (1986: 352), in his seminal article 'On the Environmental Ethics of Dao and the Ch'i', distinguished a concept of environment derived from environs, the Old French word meaning 'in circuit' or 'turning around' (Weekley 1967: 516), from a Daoist conception. The modern concept, derived from the French, he argues, suggests an external relation, devoid of either context or relationships of organic interdependence. Environment is concerned primarily with surroundings, material conditions, circumstances and a passive dead world, which builds on a philosophy of materialism, Cartesian dualism and mechanistic naturalism. Since Descartes, he further suggests (Cheng, 1986: 353), the human world has been separated from the nonhuman, and the latter studied rationally, manipulated scientifically and exploited for the maximum utility of humans. The modern framework not only rests on the human control of nature, as Cheng (2018) notes, it is bound up with concepts of political dominion and economic exploitation, and thus the control of some humans by others. The human artefact, and its meaning, has a long history in Western thought, spanning from the ancient world through the early modern period. The human artefact came to be defined in opposition to nature, as expressed in Hobbes' distinction between a state of nature and the artificial construction of the Leviathan.[5] Consistent

[5] See Bobbio (1993) for a discussion of the history of this debate. It is important to emphasize that there is no single way to interpret the human artefact in Western thought.

with Cheng's point, state-of-nature arguments have been used in historical practice to distinguish categories of humans. Africans, Native Americans and women were said to remain in the state of nature, which provided justification for acts of appropriation, enslavement or dispossession (see Snapshot 6).

The Anthropocene is the name for an era that has led to a severe imbalance between the human artefact – that is, industrialization, models of economic growth based on consumption and the development of ever more deadly weapons[6] – and nature, which has been all but destroyed. Cheng (2018) argues that to get out of 'the dangerous game of trying to conquer nature', humans need to cultivate and transform themselves, and their way of thinking, in accord with the philosophy of the Dao, making 'children, partners and friends of nature, transforming the latter into a home for human beings' rather than 'a place of machines, merchandise and profit.' The philosophy of Dao emphasizes the importance of interrelatedness and self-cultivation for becoming a more genuine human being.

Nature does not belong to a single paradigm. It isn't a 'thing', Peerenboom (1991) argues, but a conglomeration of concepts that humans apply to the world around them. Nature has been defined in terms of any number of differences, which intersect with conventions of science and culture in different ways. Within the Judeo-Christian traditions, humans are understood to have dominion over nature, endowed by a transcendent creator (see the Book of Genesis). Secular philosophers, such as Hobbes and Locke, saw nature, respectively, as a state of war which needs to be escaped, or as an inconvenient place, there to be enclosed and appropriated for use by humans. The secular construction has self-interested 'man' at the core.

The model that has given rise to the Anthropocene is at odds with the findings of more contemporary holistic science as it has emerged over the last few decades (Miller, 2006: 1–2). While religious fundamentalists, secularists, socialists and capitalists have all shared an emphasis on the primacy of the human, evolutionary studies demonstrates that humans share genetic roots with other animals, ecology has shown that human life depends on plants, trees and bacteria within an interlocking ecosystem, and environmental science that humans owe an ethical obligation to the non-human world. The anthropocentric world corresponds neither with the physical universe as understood by contemporary science nor with Daoism, but has enabled the creation of unsustainable economic cultures that threaten the extinction of life.

[6] The Anthropocene is an epoch that begins roughly with the start of the nuclear age and coincides with massive increases in population and carbon emissions, the limitation and extinction of species and with the boom in production and disposal of metals, concrete and plastic. This new epoch is partly a result of carbon-fuelled consumer capitalism which has generated a particular division between rich and poor regions and peoples. Humanity in this view is understood to be locked into systematically exploitative relationships to non-human lifeworlds.

Post-human, new materialist and indigenous literatures have sought to articulate a different ethics that incorporates an ecology of both human and nonhuman bodies (see, for example, Bennett, 2010; Coole and Frost, 2010; Cudworth and Hobden, 2011, 2013). They share much in common with Daoist thought, including an emphasis on decentring the human and the ontological parity of all life, as well as a concept of agency that is dislodged from an exclusive mooring in the individual agent. Audra Mitchell (2016), for instance, contrasts an ethics that emanates from humanity with an ethics that originates in nature, as expressed for instance by indigenous movements in the figure of Pachamama, in which no particular form of being has ontological primacy and harm and violence are understood to be directed not only at beings but at relations between them.

Decentring shifts emphasis from human control over nature to conceptualizing humans as a part of nature, thereby embedding agency not only in a larger context but in its interdependence with other life. Coole (2005: 124–42) speaks of human agentic capacities distributed across a spectrum which runs from reflexive capabilities, to the 'motor intentionality' of the body, to the nonpersonal phenomenal force field at the other end, which is shaped by the intersubjective context where action takes place. Actor-network theory seeks to explore the vital materiality that extends beyond the human, wanting to make 'things' count and enact agential power in an assemblage with human bodies, words and regulations (Latour, 1996). Latour (2004) coined the term actant, which pulls the agent into a relationship where they are mutually modifying with other actants.

Daoism provides further insight into the working of actants and 'assemblages', terms associated with Deleuze and Guattari (1987). Jane Bennet (2010), for instance, draws on the Chinese concept of *shi* (勢) to illustrate a form of agency that is attached to an assemblage that belongs to a larger force field.[7] Potentials, in this conceptualization, originate not with

[7] Notes on translation: I have added Chinese characters when unpacking the meaning of Daoist concepts. Among the variations of Chinese characters, I rely on one more traditional, and another more contemporary. Given that the concepts and/or texts in question are of ancient origin, I have placed the traditional characters within the text, and, when different, the more modern, simplified characters are added in a footnote. Characters are inserted where the meaning of a concept is unpacked rather than necessarily at first use. Phonetic transcriptions of the Chinese characters can also take multiple forms: the Chinese Pinyin and the Wade-Giles romanization. I have tried to consistently use the Chinese Pinyin except within direct quotes and with a few names, such as Lao Tzu and Sun Tzu, which are widely known in the West. The Chinese Pinyin is then provided in a footnote. English translations of chapters from the *Daodejing* are taken largely from the Ames and Hall (2003) version, which is widely viewed as one of the most authoritative on the subject.

human initiative but rather with the 'very disposition of things' (Jullien, 1995: 13). Shi, originally a word used in military strategy, refers to 'a style, energy, propensity, trajectory or elan inherent to a specific arrangement of things' (Bennett, 2010: 461). As will be discussed in more depth in the next snapshot, a good general has to read and then ride the shi of a configuration of moods, winds, historical trends and armaments. An assemblage has shi, which is vibratory, and reflects the mood or style of the whole such that parts change over time as they undergo internal alteration. Each part has autonomous emergent properties. As the parts of an assemblage change, new sets of relations and movements form, thereby transforming the whole. In the following sub-section, we focus in on the Daoist framework, and how form emerges from a formless universe, Dao, and the place of nature within this dialectic.

The Dao and nature

Daoism begins with a view of nature that stands in stark contrast to the religious/secular opposition of modernity outlined earlier. If the latter is notable for the clear boundaries that separate humans from each other, as well as nature, the state or God, the Daoist construction reveals what looks like interwoven concentric circles, or a hologram, within which the pattern of the whole is reproduced in the particular, starting with the oneness of Dao, which then becomes the twoness of yinyang, which generates further differentiation. Humans arise within nature rather than standing above and outside of it. In a hologram, information that generates the whole is encoded in each pixel, such that the whole is present in each. In a holographic universe, wholes are constituted from parts, which are themselves wholes which are constituted of parts, and so on and so on. In the radical perspectivism of Daoism, each particular element of experience is holographic; the entire field of experience is implicated within it. As Ames and Hall (2003: 18–9) state:

> This single flower has leaves and roots that take their nourishment from the environing soil and air. And the soil contains the distilled nutrients of past growth and decay that constitute the living ecological system in which all of its participants are organically interdependent. The sun enables the flower to process these nutrients, while the atmosphere that caresses the flower also nourishes and protects it. By the time we have 'cashed out' the complex conditions that conspire to produce and conserve this particular flower, one ripple after another in an ever-extending series of radical circles, we have implicated the entire cosmos within it without remainder.

Each and every distinct phenomenon is continuous with all other phenomena within a particular experience. Life requires boundaries that distinguish the internal processes of organisms from the energy and other nutrients they receive from the environment, yet life is entangled with all other life.

According to the *Daodejing* (Chapter 42), Dao(道) produced the one, the one produced the two, the two produced the three and the three produced the ten thousand things. Everything that exists belongs to the totality of things or is 'figuratively speaking' among the ten thousand things. The universe is a grand unity that began as a vast field of undifferentiated primal qi (life force), which separated into polarities of yin and yang. Yin and yang further interacted to create the diverse and complex world we live in. Nature is the physical imprint of Dao's creative power. The impenetrable paradox at the heart of this cosmology is a relationship between nothingness and being (Chai, 2018: 262). All things in the universe are constantly engaged in a cycle of dynamic transformation, break down and return to a state of qi (Meyer, 2012), the same cycle of life and thermodynamic decay discussed in Snapshot 1.

In the Daoist framework, the word cosmology must be used with some caution. The classical Greek use of cosmology, more familiar in the West, refers to a metaphysics. The cluster of terms surrounding it, such as *logos*, *theoris* or *nomos*, suggest a divinely governed universe, which allows for natural or moral laws to be intelligible to the human mind. Translated into the Judeo-Christian tradition, God is all powerful, an omnipotent other who is maker and judge (determiner) of the world. Any further creativity is secondary and derivative of the all powerful. As Ames and Hall (2003: 14) note, ' "kosmos" terminology is culturally specific' and when applied uncritically to Daoism leads to a cultural reductionism that hides significant differences.

Daoism is 'acosmotic' in that there is no a priori single ordered world or beginning.[8] As with Buddhism, there has been a tendency, as Chai (2018: 259) notes, to misappropriate Daoist thought, making it conform to a Western system. One example of this misappropriation is the identification of nature with the mystical, as antithetical to technology and subordinate to human consciousness. Instead, Chai points to a relationship between the formless and unknown and that which is manifested in form. The relationship weaves together what is seen with what is unseen. The dynamic is comparable to the plant that sprouts from a hidden root without which it would not blossom, and to the process by which pollination gives rise to

[8] The Chinese word *yuzhou* (宇宙), which is often translated as kosmos, expresses an interdependence between time and space. Rather than assuming the fixity of things in the world, as bounded and limited, particular 'things' are processual events; they are porous and flow into other particular things in the ongoing transformation that we refer to as experience (Ames and Hall, 2003: 15).

the myriad things and differentiation in nature. While the workings of Dao are mysterious, nature is not. The use of the language of mysticism in regard to nature is a romanticization which destroys what is most important to the relationship that threads the Dao and nature together while maintaining their distinctness. If Dao is the unseen root of life, Chai (2018: 261) states, nature is its limbs and branches. The Dao rests in nothingness, or emptiness, a place of quiet and stillness, while the myriad things have a place in nature. Nature is in motion, pointing to the mystery of Dao but also different from it.

The fluid and shifting boundaries between things suggests that any notion of integrity or wholeness lies not within the single thing but is instead co-creative in relation to the diverse world. Creativity arises from a back and forth that is reflexive. To be effective, this co-creation within a whole must express a relationship of respect and deference to the other, who is both separate from and a part of the self (Ames and Hall, 2003: 17). The Dao, or waymaking, is a creative process rather than a creator. Differentiation and particularity emerge from process, and the field of experience is always constituted from a contextual perspective. There is no Archimedian point or external place from which to view the totality. Change is the heart of the process and the starting point for worldmaking which pervades every aspect of the universe, as communicated in Chapter 54 of the *Daodejing* (Ames and Hall, 2003: 161):

> Cultivate it in your person,
> And the character you develop will be genuine;
> Cultivate it in your family,
> And its character will be abundant;
> Cultivate it in your village,
> And its character will be enduring;
> Cultivate it in the state,
> And its character will flourish;
> Cultivate it in the world,
> And its character will be all pervading.

Like the hologram, each part reflects the image of the whole. Humans who live in accord with this cosmology, like any other form of life, feel 'at home' in the universe.[9] According to the *Zhuangzi* (Ziporyn, 2009: 39–49), a text from the Warring States period (476–221 BCE),[10] becoming

[9] Ames and Hall (2003: 13) suggest a translation of the title *Daodejing* that emphasizes living according to the cosmology as 'feeling at home in the world'.

[10] The text is composed of 'inner chapters', which are attributed to Zhuang Zhou, and 'outer chapters', which may have come from a variety authors.

attuned to the oneness makes it possible to penetrate patterns of change with a clear head, weighing and balancing while also gaining bodily powers that transcend physical limits. The objective of meditation, 'sitting in forgetfulness' (*zuowang*, 坐忘) via 'fasting the mind' (*xinzhai*, 心齋)[11], is preparation for engagement with phenomena, and is less a spiritual exercise than nourishment of life's practice (Chai, 2018: 270). Action is entangled with and arises from the Dao and the life force, qi, making it possible to navigate one's surroundings effortlessly.

Dao and qi

The dynamic process of differentiation that arises from the polarity of yinyang is fuelled by the life force or qi. Qi, which makes all things appear, does not show itself (Hao, 2006: 456). Like emptiness, qi (氣)[12] is an abstract notion, as it does not refer to anything. As a fundamental concept of Chinese philosophy, according to Cheng (1986: 362–3), qi is expressed at different levels, in different conditions and with specific characteristics. As a level, it can represent the natural vapor of steam arising from a rice field, the human breath which infuses the body, or it may be linked to the air or energy more generally. As a condition, it is alternatively a vital energy with no form and transformative in giving rise to the form of all things in the universe. As a characteristic, it is omnipresent, resonating among all things in the universe, varying as it flows from differences of kind and structure. Far from arbitrary, it brings order, structure and form that contributes to the proportionate sustenance of life and gives rise to laws and patterns of nature. Qi can take the form of yang or yin within a whole, yet potentially expresses the dominance of one or the other as things change and transform. There is nothing in nature that is not qi. It is the life force, the breath of life, the emptiness of emptiness, from which two fundamental concepts of *wu* (no-things) and *you* ('things'), formlessness and form, arises. As in the plant example, qi is the activation by which the unseen manifests itself in the seen. As I will elaborate on shortly, the translation of wu and you in terms of thing or no-thing does not capture the subtlety of the distinction; however, for now, their translation as thing and no-thing will suffice.

Non-presence/presence

Complementarity in quantum mechanics means that the fundamental constituents of being are no longer things distinguished from other things,

---

11 心斋.

12 气.

as in classical physics, but rather a particle-wave relationship by which any one particular may have a material expression which, to be understood in its fullness, must account for a polar other, for example the hidden wave. The relationship is not static but dynamic. 'Particular' shares a root with 'particle', which refers to the 'small part'. Particular pertains 'to part but not all', or special rather than general. 'The small part' suggests a relationship to a larger whole, but the impermanence and dynamism of this relationship is far different in the absence of predetermined patterns or first principles such as are assumed by the metaphor of mechanism in Newtonian physics. Objective reality, in the Daoist and quantum reading, is objectless. In an ever-changing world, an otherwise dynamic movement may temporarily freeze a particular pattern of difference, but the pattern will eventually be subject to change and transformation.

Quantum complementarity has a parallel in the Daoist yinyang. The term yinyang embodies a wide range of meanings, many of which may be in play at the same time. Yinyang, as two interrelated and interdependent parts, has many expressions: it can be the key to fostering life or a normative model of balance, harmony and sustainability; a correlative mode of thought or cosmology; and/or a constellation of beliefs and practices relating to a range of activities from philosophy to healthcare to battle or a way of life (Wang, 2012: 2–6). As Robin Wang (2012: 5) states, 'yinyang places human flourishing within a rich and deep context involving the interrelatedness of the cosmos and human beings'. The term represents a paradigm for developing strategy for effective functioning in the world, which can be applied to many aspects of life.

In Chinese, yinyang is a single word (陰陽),[13] which highlights the unity from which the polar opposites emerge. Yinyang can be distinguished from the usual English representation as yin *and* yang (Wang, 2012: 6). When the parts are separated by the conjunctive 'and', they are prior to the whole. Yinyang, by contrast, makes the separation a function of the whole, expressed in a single word yet characterized by dynamic transformation. In common use, yin is associated with stillness, weakness, femininity, mystery, while yang is associated with strength, action and masculinity. The oppositions, when presented in this way, easily fall into static and hierarchical categories that become attached to flesh and blood women and men, or come to embody what it means to be a woman or a man. In this state, the dynamism of the emptiness/form relationship, as well as a universe of continuous change, gets lost. What is empty or full can only be defined relationally (Wang, 2012: 152).

13 阴阳.

Figure 2: Yinyang symbol and nature

Source: Art by Milan Loiacono, Pepperdean Graphic

Yin and yang are derived from an example in nature, that is, the sunny and shady sides of a mountain (Lao Tzu, 1963: 125). Each perspective provides a different angle on the same phenomenon. The angle of the sun shifts over the course of any one day (time). The observer moves to different locations around or on the mountain (space). There is no one place from which the whole can be seen in its fullness. From any one position the brightness or the shade will be more dominant, as shown in the artistic rendition of the yinyang symbol in Figure 2. Neither is static. Both are continuously changing. Yin and yang together constitute knowledge of the mountain, which changes depending on the position of the observer.

The relationship between the sunny and the dark side of the mountain is first and foremost one of presence and of that which can be seen, on the one hand, and that which is not present and thus not seen, on the other. Yin is the unseen and yang is the seen. As such, the Chinese categories of you ('things', 有) and wu ('no-thing', 無),[14] as suggested earlier, do not map onto being and non-being (Wang, 2012) or essence and appearance (Chen and Schonfeld, 2011: 199), distinctions that are more common in Western philosophy. They instead refer to presence and non-presence (Hall, 2001: 250; Wang, 2012: 17). Being/non-being or reality/appearance express a relationship between that which is or exists as a subject or object, and which therefore expresses truth, and that which is not (for example. mere appearance or absence). Wu and you have a different connotation. Being

[14] 无.

does, of course, constitute a presence, and non-being is a non-presence, but the significance changes if one moves from a separatist ontology to Daoist holism in which all things are interrelated. Truth and appearance refer to a separation between that which is real and that which is not real (mere appearance). They are different in kind, and truth has ontological priority over appearance. By contrast, presence and non-presence are entangled phenomena in the Daoist conception and have ontological parity, which is a difference of degree.

Daoism reveals the self-generating dynamic of the process of becoming. The dynamic relationship between opposites can refer to any number of relationships, not least gender as well as human nature and emotion. The key points in highlighting the entanglement of present and non-present are as follows. First, yinyang is a relationship of presence and non-presence within a whole. Second, as in the discussion of emptiness and dependent origination in Buddhism, present and non-present are intertwined rather than separate, and neither has ontological primacy. Just as the wave function as potential collapses into a particle, yang – that which is present – emerges from yin – the non-present. Third, that which is present and non-present is continuously changing; that which occupies the yin (non-present) in one context may occupy the yang (present) in another (Wang, 2012: 6). Finally, both action and strategy arise from navigation of the dialectical relationship between yin and yang, with the movement of each towards its opposite.

The cycle of nature involves a reversal of opposites. That which is new becomes old; that which is beautiful in time becomes ugly; that which is light returns to the shade. As Lao Tzu says in Chapter 2 of *the Daodejing* (Ames and Hall 2003: 80):

> As soon as everyone in the world knows that the beautiful are beautiful,
> There is already ugliness.
> As soon as everyone knows the able,
> There is ineptness
> Determinacy (you) and indeterminacy (wu) give rise to each other.
> Difficult and easy complement each other
> Long and short set each other off,
> High and low complete each other,
> Refined notes and raw sounds harmonize (he) with each other,
> And before and after lend sequence to each other –
> This is really how it all works.
> It is for this reason that sages keep to service that does not entail coercion (wuwei)
> And disseminate teachings that go beyond what can be said.

As in Buddhism, Daoism contains no notion of intrinsic being or static human nature, and thus no intrinsic beauty or ugliness, good or evil.

An important point, as suggested by the ordering of yinyang, is that yin, or that which is non-present, and thus unseen, comes first. Form emerges from formlessness. In so far as this pattern is reflected in nature or is patterned on nature, it is akin to the seed buried deep in the ground from which the plant emerges. A similar idea is expressed in the root-branch metaphor that appears in Daoist military works (Meyer, 2012: 32). The root is the hidden non-presence from which the multitude of branches emerge. The significance of the relationship between presence and non-presence pervades everything, down to human action and strategy.

Yinyang is about bringing the hidden background factors into awareness (Wang, 2012: 147). As Ames and Hall (2003: 34–5) note, the ability to foreground the relational character of a particular matrix of events, and the field of conditions from which it arises, creates a capacity to see where a situation comes from and to anticipate where it is going, as well as to make minor alterations at an early stage of a process, which may have cascading implications for the outcome. This contrasts with an approach that isolates things and makes judgements on the basis of the immediate phase in which they are observed, thereby reifying its objectivity at a time when it may still be evolving.

The preference for the formless is particularly evident in the frequent use of a water metaphor by both Lao Tzu and Sun Tzu, which again relies on a pattern from nature. Water is formless, having no shape of its own. It changes and adapts as it rushes down a slope or courses through a river. Its shape is defined by that which contains it (Lui, 2001). The container itself is empty except for the water it holds, while the water changes form to fit the boundary provided by the container. Water can be soft and gentle or it can destroy (Lee et al, 2009: 75). It is a fierce enemy in a flood, or, in its persistence over time, can wear away the hardest of rocks (Lao Tzu, 1963: 85). Water is formless but always engaging with and shaped by form. Water sinks to the lowest places, but, precisely because it is skilful at staying low, it is able to spread and connect with the multitude of waterways, which is the source of its power (Lao Tzu, 1963: 73).

What is crucial in the formlessness/form relationship is balance between the two parts. As Wang (2012: 146) argues, the need to tend to yang when yin is strong and yin when yang is strong is a central component of strategy. Balance is less a matter of equal weight, as in the mechanism of a scale, than of the resonance of yin with yang or the mutual movement of polar opposites. Chapter 24 of *Zhuangzi* (Ziporyn, 2009: 103) recounts the story of a disciple who wanted to get the Dao because this would enable him to make a pot of water boil during the winter or to make it freeze to make ice in summer. As boiling water is needed in winter and ice counteracts the heat of summer, the objective would seem on the surface to be consistent with yinyang. But the master was not convinced and responded that the disciple was using yang to evoke yang and yin to evoke yin. To illustrate how the getting of Dao is different, the master placed zithers in two different rooms, demonstrating that when a note was plucked on one zither, it would resonate with the other. Further, after changing the tuning of one string so that it matched no proper notes, he showed that twenty-five strings on the zither would resonate with the plucking of one, revealing it to be the master of all the others, by stimulating them to resonate along with it. Zhuangzi here anticipates quantum discoveries such as the forced-resonance tuning fork experiment, in which a vibrating tuning fork forces a stationary tuning fork into resonant vibration when placed beside it (Ooe, 2014).

As Ling (2014: 95) notes, the emphasis on balance compels a consideration of its opposite, that is, imbalance, as well. When each side of a polarity is paired with itself, rather than its opposite, the result will be disaster. Within Daoism, floods or pestilence are understood to be a response to an imbalance in nature. In a situation of imbalance, when one extreme is reached, nature will shift to the other extreme. In ancient texts this reversal is associated with a concept of cosmic resonance or *ganying* (感應).[15] Cosmic resonance is indebted to neither spiritualism nor an angry creator. Ganying is a philosophy of correlative thinking that posits the intrinsic relatedness between all things, from cosmic forces to all species and natural objects as well as aspects of human life such as government (Lai, 2007: 327). Because all things are made of qi and interconnected, stimulating any one thing in the universe creates a simultaneous response in other objects or people. Ganying, as a stimulus-response mechanism, represents, as Scharf (2002: 83) notes, the 'sympathetic resonance of objects belonging to the same class, which resonate with each other just as two identically tuned strings on a pair of zithers.' Resonance is as applicable to the relationship between rulers and subjects as to humans and the universe or nature and world. Correlative cosmology was a rational and empirical approach based on the embeddedness of all within the Dao (Meyer, 2012: 27).

The differentiation of the ten thousand things is biodiversity, with each separate species using intelligence to navigate through life. Humans are a part of this diversity, yet themselves generate human artefacts. These forms may be more or less attuned to nature, and when they are not, contribute to imbalance. To get to the Dao is to strip back the human construct and to act from within a pattern of nature, in tune with its resonances. Both Lao Tzu and Zhuangzi suggest the need to undo human fabrications in order to reveal the power of the Dao contained within them. Behind the concepts all

15 感应.

belong to the single relatedness of the Dao. Form is entangled with hidden formlessness. The objective is to unravel this contradictory construction, to reveal the illusion.

Hao Changhi (2006: 454) explores two stages in Lao Tzu's process of deconstruction, starting with the negative task of showing the relativity of the human construction and the positive one of establishing the primary philosophical concepts, such as wu (no-thing) and you (thing-ness). In a second reduction, the philosophical concepts are deconstructed in order to demonstrate that Dao is beyond human language. Restated in terms of the Buddhist concepts, the first reduction establishes the relationship between dependent origination and emptiness, revealing the reliance of the former on human artefact; the second reduction deconstructs emptiness itself, revealing that it too is empty. The process of reduction outlined by Hao suggests a return of the human to an alignment with nature.

The relationship of yinyang is not one of either/or, whether the oppositions constitute emptiness and form, non-presence and presence, light and dark, boiling and cold. Yinyang is a relationship of either/and, constituted by the resonance between polar opposites. The concern in looking at the contemporary environment regards the relationship between the human construct and the human place in nature, which is a necessary background for placing wuwei (actionless action) in the realm of action rather than passivity. The relationship between unseen and seen is deeply entangled rather than being a hierarchy constructed on separation. In quantum entanglement, when *x* moves up, *y* moves down; they move in sync, in a relationship, and the movement constitutes a resonance and vibration that is non-local. The Daoist yin and yang express the same phenomena. Nature expresses the balance and dynamic emergence of form from formlessness: the biodiversity of the multitude of the ten thousand things, infused with the life force of qi. The potential for imbalance increases as the human artefact, a conventional construct, is imposed on a world, in particular if the artefact is in conflict with the rhythms of nature or seeks its control. Against this backdrop, the concept of wuwei or actionless action takes on a particular meaning.

Wuwei

Wuwei (無為)[16] is one of a cluster of related terms used throughout the *Daodejing*. The concept appears twelve times in ten chapters (Liu, 2001: 318). Wu, which refers to non-presence or nothingness, appears 101 times in the *Daodejing* (Wang, 2012: 56). Wei has a general meaning of doing, making

[16] 无为.

and acting, that is, of a presence. A frequent translation of wuwei is 'no-action' or 'taking no action'. However, the translation of wuwei as no-action removes the contradiction, thereby reinforcing a connotation of passivity or of not acting at all. Wuwei as no-action is contrary to its spirit and connotation in the *Daodejing*, which maintains the dialectical tension between polar opposites of action and non-action. Both sides of a contradiction are contained within the whole, a possibility that is accepted within Chinese thought. A straightforward translation of actionless action as non-action is an oversimplification that misses the spirit of the concept, which resides in the dialectical relationship between yang (action as presence) and yin (without the presence of action).

The Daoist concept of actionless action, wuwei, is often misinterpreted as non-action, mysticism, passivity and/or fatalism (Nelson, 2009: 303). To say that wuwei is in essence passive is different, however, than understanding non-action as a yin form. Yin, in its hiddenness, constitutes, interpenetrates and embodies potential action. The dynamic nature of the figure/background relationship is central to the strategic potential of actionless action. Attentiveness to the background as the unseen, and thus to contradiction, makes it possible to work in accord with a quantum reality rather than in opposition to it. In the Daoist conception, the key to strategy is to rely on the potentials that emerge from within a context or situation and to flow naturally with them as they evolve (see also Jullien, 2004: 20; Wang, 2012: 123, 137; Yuen, 2014: 78). Effective and effortless action will then unfold in accordance with the contextual pattern, rather than attempting to control or resist it. The actant becomes a participant in a process, both moving with the change and, at the same time, strategically shaping it. The strategic aspects of Daoism will be developed in the next snapshot. Here the central point is that the actant is internal to the context rather than representing or manipulating a 'reality' outside the self (Wang, 2012: 140). She participates from within a set of relationships that are interpenetrated. The interdependence of parts suggests a form of action that departs from the more familiar model of autonomous rational action. Movement from within a relational web may appear to be more constrained than from a position of separation and autonomy, but this does not mean an absence of differentiation.

Wuwei conceives of action as emerging from a relational context as the self becomes action and becomes a participant in the process of differentiation as she engages with the world and the other. The claim that self becomes action highlights the emptiness of both, given the mutual constitution of self and others within a relational whole. Karen Barad (2007: 178) has used the term intra-action to refer to the process of mutual constitution. The more conventional use of inter-action, situated in a classical ontology of separation, leaves the separate interacting parts intact. Intra-action, by contrast, is the process of differentiation by which *selfother* as a relational phenomenon comes into being. Her argument grows from a critique of the logic of representation.

Representation is referential. It involves an imposition of meaning in the naming of objects, events, states or individuals as types or kinds. A representation is assumed to be more or less true vis-à-vis the world. What Barad refers to as intra-action would, from a Daoist perspective, be seen to arise from a relationship of deference, in contrast to representation. Hall (2001: 247) argues that the deferential eschews naming or classification. It is necessarily circuitous and yields to the subject or object of relationship. References are established vis-à-vis a common standard, that is, based on formally defined concepts to identify members of a class of items. Mutual deference, by contrast, is patterned on nature. The latter claim raises two issues.

First, as any observation of nature will reveal, a pattern of deference is not equivalent to the absence of violence. Violence is a part of the pattern of life and death, of survival in nature. The violence of nature is not, however, mobilized in the pursuit of profit or death for its own sake. It is less that violence is absent or present than that entangled engagement provides an incentive to minimize damage to the other. The ontological shift from separateness to relationality implies an epistemological shift, to be elaborated in what follows. Human representation as a mirror of objects in the world, for the purpose of capturing 'truth', is replaced by the mutual mirroring of actants engaged in practices of mutual deference.

Second, mutual deference, as already suggested, is not, by definition, respect in the positive sense but rather an acknowledgement. It arises from a relationship between self and environment that recognizes their mutual implication and the need for balance between the two. Life is entangled, so any defence of self needs to also consider the costs of destroying a shared habitat upon which all life, in its diversity, depends. The Daoist concept of deference suggests a mirroring that engages the world in its impermanence and particularity (Hall, 2001: 254). Rather than imposing form, the language of deference is necessarily circuitous – it yields to the perspective of things that are to be acknowledged and appreciated.

Deference is a language of presence and making present rather than a representation of the essence of a thing or event, and thus suggests an absence of absolute ontological foundations. The engagement takes place within a real world but not a world of objects. The reality is one of constant change that precludes 'a final inventory of its contents', as such an inventory would miss the objectivity of the world in all its complexity (Hall, 2001: 260). Newtonian science builds on an ontology of being and non-being, matter and nothingness. Daoism, as well as quantum science, is by contrast concerned with presence (you) and non-presence (wu). The *Daodejing* gives preference to the latter (Hall, 2001: 251). The difference is subtle and often leads to the misguided assumption that Daoist thinkers prefer the void or non-existence over being. However, the absence of a notion of being as existence only means that there is no notion of being as ontological ground in classical Chinese philosophy, and thus no need for a metaphysical contrast between being and non-being. The language of presence is a language of making present not the essence of a thing or event but the thing or event and its process of becoming (Ames and Hall, 2003).

What might it mean for humans to engage deferentially with the environment, as a part of it, rather than standing apart from or over it? The conversation starts within, with the stilling of the inner narrative, which is preoccupied with self and its own emotions and assumptions. A more peaceful equilibrium within makes it easier to hear the sounds of the background environment, for example the breaking of the branch, the howls of an animal, the approach of an attacker. But the listening can extend further. Following the enforced stillness and tranquility that accompanied the nearly global lockdown during the first wave of the COVID-19 pandemic, many people noted not only the freshness of the air but the sounds of animals and birds precisely because the environment was not, as it usually is, filled with the smells and sounds of traffic, whether on land or in the air, or with the range of human practices that typically cover over the sounds of nature or even our ability to see it. From an apartment building in London, one resident commented that he had never before noticed the tree that was visible from his window. Acknowledging nature, and deference to it, starts with 'seeing', of making the non-present present.

Human practices often arise from the imposition of meaning on the environment based solely on human representation for human ends. One might venture further to observe the consequences of this emphasis on representation rather than deference with the emergence of the Anthropocene or the gradual destruction of nature and the near completion of the human project. The ancient Chinese concept of ganying provides a correlative framework for beginning to see the increasingly frequent weather events, as well as the emergence of the pandemic, in the context of nature demanding to speak for itself and be heard, precisely because humans have refused to listen until forced to do so by a deadly virus. Again, this is not to suggest that there are supernatural forces at work, such as an angry god, but very natural forces, evident within nature, that reflect a process by which patterns, when they reach their extreme, revert to their opposite, thereby mirroring the natural cycle of destruction and rebirth.

In the context of the reversals brought about by the pandemic, the world began to look like a different place. At least for a time, the selfishness of the old normal came more clearly into view. Isolation gave rise to an outpouring of compassion. The ugliness of practices that pollute the earth became more unmistakable against the backdrop of a moment of freedom from these pollutants. The importance of human connectedness and touch became painfully evident in the context of self-isolation and physical distancing. Through the non-local medium of social media, populations nonetheless experienced a heightened sense of connectivity, which was all the more pronounced because of the absence of physical proximity. The example reveals two patterns that provide the experiential foundation of wuwei, that is, the interdependence of opposites and the reversibility of opposites (Liu, 2001: 325). The polar construction of yinyang means that life and death can no more be treated in exclusion than good and evil or beauty and ugliness. Each is known through the contrast with its opposite. The opposites are entangled in their complementarity even while only one side of the polarity is present and seen at any one point in time.

From this picture of self, situated in a natural environment, the meaning of wuwei begins to emerge, although it differs somewhat depending on whether it is understood as a methodological principle, a guide to political action or a principle of governance. Emphasis on the dialectic of actionless action in itself suggests multiple potentials, as well as a greater opportunity for misunderstanding. The relationship between wuwei and nature is clear. Wuwei in the *Daodejing* (Chapter 2) is aligned with nature and the origin of the universe and all things within it. But actionless action, no less than mind/no-mind, rests on paradox and contradiction.

Method, mind and action

The shift from a logic of representation to one of deference is methodological. An epistemology of representation gives way to an experience and method of conversation, which transforms the epistemological and ontological ground of engagement with other and world. Representation imposes form on a reality that is believed to exist independently. The primary concern is whether the representation is more or less true. Deference is instead a yielding and action within a transformative relationship which takes the other into account. The meaning of what is said and the process of conversation is more important than establishing what is true. The constitution of knowledge as an 'inventory of facts' gives way to knowledge and insight that arises from looking to new horizons or bringing different perspectives into conversation (Hao, 2006), a shift from 'truth' regarding what 'is' to new potentials.

The *Daodejing* is about differentiation or how it is that a deeply entangled and formless universe finds expression in the diversity of the ten thousand things. The claim that diversity rests on a degree of complementarity and coordination does not deny the status of any one individuated phenomenon it itself, which is a function of that phenomenon's spontaneity (*ziran*,

自然). Arthur Waley (1958: 174) translates ziran as 'self-so', which in his argument is the 'unconditioned' or the 'what is so of itself'. The self here is non-reified, reflexive and a self in context. The self is spontaneous in so far as it is not subject to conformist and universally binding norms (Hansen, 1992: 213–4), but also because it expresses an ethical judgement that works beyond the boundaries of convention (Lai, 2007: 336). Ethical action rests on a dialectical relationship and balance between normative structures as human artefacts and the spontaneity of the individuated action. While normative structures may enforce conformity, and prescribe a 'natural path', the immediacy of a rapidly changing situation and interdependence with others means that the artefact cannot be static or universally binding. It must evolve through the mutual engagement of individual specificity in the larger context of Dao (Lai, 2007: 327). One transforms along with the rhythm of nature and the world. Ziran and wuwei taken together emphasize the mutually responsive and spontaneous nature of relational interaction.

The different concepts are like pieces in a puzzle that construct a picture of how things in nature, through the process of differentiation, work in complementarity and coordination to produce the world where everything is 'in the soup' (Ling, 2013: 99). 'In the soup' points to the etymology of the Chinese character for harmony (*hexie*, 和諧),[17] which is culinary. Harmony in this sense is the art of combining or blending two or more ingredients such that that they enhance one another without losing their distinct flavour. Harmony is less a category of general law than the aesthetic composition of the contextual. From the contextualized disposition, ziran (that is, the spontaneity of action) arises from the unfolding of natural processes when they are left free of interference. This dispositional spontaneity arises both from a pattern of nature, and one's position within it, as, stated in Chapter 25 of the *Daodejing* (Ames and Hall, 2003: 114): 'Human beings emulate the earth, the earth emulates the heavens, the heavens emulate way-making [or the Dao] and way making emulates what is spontaneously so (ziran).'

Movement from formlessness to form is itself spontaneous, yielding to a relationship of complementarity and coordination with others within a whole. The 'thing' becomes itself in its uniqueness, and thus contributes to and shapes the whole. In this respect, humans aren't simply agents manipulating the world through representational knowledge but participants in the process of creating the world (Wang, 2012: 140). Dao is the ontological source from which all things derive their existence, and *de* (德) is the particular instantiation, the individualizing factor and embodiment which gives things their determinate features (Chan, 1963: 11) and the power and

[17] 和谐.

efficacy that comes from harmonizing with the Dao (Wang, 2012: 129). The first pattern mentioned, that is, the interdependence of oppositions, highlights a relationship of deference, and the second reinforces that none of this rests on essence. Positions and perspectives change. That which is present and that which is non-present, and the relationship between them, changes. Action necessarily begins with an awareness of the impermanence of any particular identity or relational pattern. To act spontaneously thus means to be able to respond with immediacy, without reference to established norms or principles and with clarity from whatever place one occupies at a particular time, and to do so with some attention to how harm to the other will impact on the self. Spontaneity thus contrasts with the imposition of form through norms and conventions.

To return to the pandemic, the economic response by a Conservative government in the UK to the need for a society-wide lockdown demonstrates aspects of wuwei and how it works in a situation of uncertainty. Everything changed almost overnight with the increasing threat posed by the spread of COVID-19 and the need to shield the most vulnerable in society and keep the National Health Service from collapsing. A Conservative government with a legacy of austerity going back more than a decade – a legacy that had exacerbated inequalities – reversed their position and implemented one of the most generous economic policies in Europe in order to provide the salaries of workers on furlough so that they would not be fired. The government acted spontaneously, creating economic policies that were in direct opposition to those associated with the party historically. In so doing, they contravened the party's own existing norms and conventions. They further reverted from one extreme to another in an attempt to redress the imbalance exposed by the virus. The reversal was also in part necessitated by the legacy of austerity, as a result of which the NHS and portions of society entered into the crisis with lower than normal resources.

The Conservative leaders might, in this one instance, be said to be following the rhythm of nature in a situation of dramatic change and uncertainty, and to be doing what the context required. Part of this reversal involved beginning to rethink the relationship between individual and society. An earlier Conservative prime minister, Margaret Thatcher, stated that there is no such thing as society, only individuals. Prime Minister Boris Johnson, who himself became ill from the virus, proclaimed that society does indeed exist, which provided a framework for individuals to conceive of themselves within a larger whole.[18] Addressing the pandemic required that people become conscious of the impact of their own behaviour on others, that is,

[18] Johnson's statement resonates with a more organic conservatism that sees society as a network of relationships that rely on duty, traditional values and established institutions.

that some deference was shown in acts of social distancing and the wearing of face masks, even if one was not oneself in a vulnerable category vis-à-vis the disease. The loss of life in various countries was largely a function of the ability of people to exercise this deference toward others. During the early waves of the pandemic, the most highly individualistic cultures, such as the UK and US, suffered the highest death tolls. Even as death tolls in both locations spiralled out of control, protestors resisted restrictions on their movement based on arguments that their human rights were being violated.

Yinyang grounds correct human action in the world and universe and thus requires a strong awareness of one's surroundings, of relationality as well as rationality (see Qin, 2016). Deference is an emotional relationship whereby the point of reference is the other as constituted within the same whole as the self. Contrary to the heavy emphasis on rationality, to the exclusion of emotion, which has characterized the dominant traditions of thought and science in the West, Daoism has a method of heart/mind that involves both thinking and feeling (*xinshu*, 心術)[19] and the proper cultivation of emotions and desires so that they are in balance, which involves managing qi and nourishing the heart/mind (Wang, 2012: 131). Heart and mind interpenetrate each other, rather than being mutually exclusive. Human nature (yang) and emotions (yin) are both good and bad, not either/or (Wang, 2012: 132). Internal balance and calm are essential to balanced action. As such, one stands in the midst of change, following the water's current rather than paddling upstream, working with changes rather than against. Standing within a whole or a self-generating system that is undergoing continuous change is a different positioning than that of the agent-structure relationship. Structure presumes an ontological ground, whether of mechanism or of other forms of order. As King (2006: 54) notes, lack of structure (or emptiness) provides the true strength of non-action. Deference is a relationship to the other which includes a place for emotions. The objective of balance is not purely a function of the instrumental ends and interests of the self and other selves. The heart/mind of the one with whom one engages must also be taken into account.

One half of the *Daodejing* deals with the practices of government, including wuwei as a key component of successful governance. The sage (or wise ruler) is situated within a pattern of nature, where perspective and spontaneity, as well as emotional balance, are crucial for deciphering the proper course and timing within it. It is not only about action itself but about being able to recognize the appropriate timing and situation. Actionless action is context bound. Any one expression is only applicable to a specific time and place, and thus cannot be understood on the basis of universal rules

[19] 心术.

or standards. Actionless action is a function of the Dao which cannot be grasped in full even while it pervades all things. Non-action does not refer to an absence of action but rather a course that brings about complete and effective results precisely because it navigates rather than trying to control or rule the impermanence. As stated in Chapter 29 of the *Daodejing* (Ames and Hall, 2003: 122):

> If someone wants to rule the world, and goes about trying to do so,
> I foresee that they simply will not succeed
> The world is a sacred vessel,
> And is not something that can be ruled,
> Those who would rule it ruin it;
> Those who would control it lose it.

Too much action or intervention may produce opposite outcomes to those which are desired. Actionless action, far from an absence of action, is a state of mind to be maintained while acting (that is, emptiness) (King, 2015: 55). This state of mind works from within the contradictions rather than attempting to resolve them.

Wuwei has roots in both Confucianism and Daoism (Ames, 1981). While Daoism has informed Confucian applications to politics, Confucian interpretations have placed greater emphasis on the human artefact, that is, skill development (Slingerland, 2000) or self-cultivation, rather than on the patterning of nature (Ames, 1981). In practice, Confucianism has often reinforced a message of conformity accompanied by tight governmental control over populations and more hierarchical authoritarian forms, which is a common stereotype of Asian politics. The message of the *Daodejing* is somewhat different. On the one hand it stresses the need to avoid strong desires arising from greed and revenge, given that they are conducive to imbalance. On the other hand, emptiness is associated with flexibility and equilibrium rather than conformity or passivity. Emptiness of mind is not only a virtue of people but also of rulers. Chapter 37 of the *Daodejing* (Ames and Hall, 2003: 134), for instances, states, in regard to nobles and kings:

> Realigned with this nameless scrap of unworked wood,
> They would leave off desiring.
> In not desiring, they would achieve equilibrium,
> And all the world would be properly ordered of its own accord.

As in Buddhism, desire for profit, greed and revenge are the source of the problem. Acting on desires rather than from a stance of non-contention leads to conflict. Indeed, one common interpretation of wuwei at the level of government is that of non-interference (Hao, 2006: 54). Non-interference in Daoism, not unlike its more liberal definition, suggests allowing a system to re-establish its own balance without intervention, as the latter may be counterproductive. The systems orientation here is key to the distinction between non-interference and simply 'doing nothing', which would be a problematic orientation for any government, at least over the long-term.

In the political world, wuwei usually means that rulers should not interfere in the life of the people or the order of the world. As such wuwei sees human action as an interference in processes that would otherwise follow their own course. But wuwei can also refer to an 'undoing'. In so far as wuwei relates to undoing or reduction, as discussed by Hao Changchi (2006), it involves a return to a more natural state, such that a matter can manifest itself anew, like the seed that gives rise to the plant or the Dao emerging from the emptiness of mind. Actionless action, far from non-action, suggests movement which is effortless because propelled along by the current of change.

Doing or acting with the Dao, often referred to as the Way, one is able to realize the potentials of a situation. Non-interference is less about doing nothing than acting in a way that minimizes risks and maximizes potential precisely because it arises from the pattern of nature. Physical terrain is a pattern which, when combined with the centrality of water, provides a potent example of wuwei. *The Mengzi*, Chapter 3, recounts the story of Yu the Great, the legendary founder of the Xia Dynasty (2070–1600 BCE) who mastered flood control techniques to tame rivers and lakes (see Wang, 2012: 142).[20] Yu's father, Gun, had been asked by King Yao to tame raging flood waters, and over nine years built dikes across the land in the hope of containing the waters. However, the dikes collapsed during a period of heavy flooding, and the project failed. Yu, who learned from his father's mistakes, took a different approach to managing floods. Rather than using dikes to stop and control the water, he dredged new river channels to redirect the flow of water, thereby flowing with rather than resisting its natural tendencies. The channels then provided outlets for the torrential waters, as well as irrigation conduits for distant farmlands, thereby successfully solving the flooding problem (Van Norden, 2008: 111, 168). As Wang (2012: 142) notes, Yu's method was a metaphor for flowing along in attunement with the terrain in order to get things done by using 'the way of water' and 'enacting what was without work'.

Liu (2001: 317) argues that human actions have caused and exacerbated the environmental crisis in post-industrial societies and are thus contrary to

[20] *Mengzi* 孟子 'Master Meng' is a collection of stories of the Confucian philosopher Mencius (385–304 or 372–289 BCE) that regards his conversations with rulers, disciples and adversaries.

wuwei. They have been broad in scale and rapid in expansion, which can be contrasted with life prior to industrialization when people acted naturally and slowly and thus without ruining their environment. Indeed, a few hundred years of industrial activity has brought on climate change, while human communities lived sustainably, if more simply, for millennia. Industrialization and commercialization rest on patterns of economic development that give rise to prosperity – and poverty. Prosperity rests on high consumption, which is generated by marketing that stimulates human desire. The spread of these practices has been underpinned by the religious/secular dualism discussed earlier. The Judeo-Christian view of humans as creatures of God bestowed with God's will and a right to manage nature has been reinforced by scientific confidence and a desire to make use of the environment. Both rely on a separation between humans and nature. More recently, in the face of climate change deniers, science has played a leading role in establishing the damage done by human action to the environment. The environmental movement has emphasized the need to look to the science.

The purpose in setting out this distinction is not to deny the useful role that science can play but rather to question what is assumed about science, the nature of reality and control. The objective is to shift thinking to new horizons in the context of a continuously changing reality. In 2021, in the midst of a further wave of the pandemic, a Cambridge economics professor, Partha Dasgupta, published a 600-page review, commissioned by the UK Treasury, which represented the first time that a national finance ministry had authorized a full assessment of the economic importance of nature (Elliot and Carrington, 2021). The report claimed that the cost of prosperity was devastating to the ecosystem that provides food, water and clean air to humanity, and that long-term prosperity required rebalancing the human demands on nature with its capacity to supply them. Nature has been a blind spot in economics, and it is dangerous to continue to ignore it. Several themes are evident here. One is the suggestion of gross imbalance and the need to rebalance. A second is the need to bring nature, which has been invisible in economic thinking, to the forefront of what is valued. Profit cannot be the only value of the human economy. The habitat in which we live, and the larger environment provided by nature, has a crucial place.

One further source of imbalance and incompleteness arises as classificatory schemes clash in practice, which can be the case when scientific models or other more yang-like forms of action clash with indigenous forms of life. The materialist and utilitarian approach of science to climate change stands in contrast to more localized symbolic values attached by local communities to, for instance, glaciers, which highlights how intersubjective structures of meaning, including those of science, shape what is seen and determine whose practices matter (Wiener, 2018) and thus how particular challenges are addressed (see, for instance, Byg and Salick, 2009; Adger et al, 2011;

Allison, 2015). For example, several decades ago Greenpeace activists were successful in pursuing a ban on seal hunting. The ban had the unintentional consequence of destroying the Inuit way of life, which is dependent on seal hunting, thereby creating other social problems. The environmental movement, as a form of struggle, sought to improve the environment, but with unintended consequences, and in a manner contrary to wuwei. The example juxtaposes two expressions of yang qi, that is, the gratuitous exploitation of seals in pursuit of profit and the struggle of environmental movements to bring about change with the unexpected consequence of destroying sustainable patterns of seal hunting by indigenous populations who lived in resonance with nature (Liu, 2001). Thirty-some years later, northern indigenous tribes continue to feel suspicious of Greenpeace and have been hesitant to join with them in an alliance to ban arctic oil drilling (Pang-White, 2016: 291). A further example in Australia involved a project to remove all feral cats from Macquarie Island to save the native seabirds, which resulted in an overpopulation of rabbits that then destroyed the vegetation that the seabirds depend on for shelter.

While humanity has been aware of the threat of environmental destruction for decades, day-to-day decisions have tended to be dominated by other, primarily economic, concerns in conformity with human patterns at the expense of nature's patterns. This changed suddenly in the context of the COVID-19 pandemic as not only humans but their polluting went into lockdown. As a consequence, nature began to be present, seen and appreciated in industrialized economies. As the context changed, there was not only more opportunity to work with nature but an aesthetic as well as economic incentive to do so. A potential for action was offered up by the context, and with it a potential to either work effortlessly with the environment or continue practices that harm it. During the first wave, public transport became so dangerous that people began buying bicycles and, where possible, pedalled to work. Public authorities considered a greater investment in bike paths. Concerns about food supply as well as mental health led to a boom in home gardening. The economy had to shift to meet the emergency. As some businesses were lost, new areas opened up, some of which might contribute to a more sustainable future. The result was an opportunity in the face of disaster to adapt the human artefact, and technology, to work with rather than against the patterns of nature.

But again, the impacts of this new context were not experienced equally around the globe. More industrialized economies benefited from clean air and the presence of nature, whether wealthy or poor. Evidence of a change in the relationship between nature and human society looked different in places where humans, who may have already been closer to nature were forced, due to the lack of employment and access to food that came with the lockdown, to engage in activities such as poaching that would be detrimental to nature. For instance, Lindsey et al (2020) argued that 'the combined effects of reduced conservation efforts and increased poverty could create a positive feedback loop where intensified reliance on natural resources spurs human encroachment into natural habitats, increases exposure to and consumption of wild animals, and amplifies future pandemic risks.'

Conclusion

The strategic dimensions of wuwei arise from a holistic cosmology, within which an oppositional dynamic gives rise to a self-generating process of differentiation, fuelled by the life force, which permeates everything in the universe, including the human body. Self-generation suggests the existence of a self who controls the process of their own development, but, as in the last two snapshots, this notion of self must be eschewed. The self is relational, and action is less about intervention or control than identifying the opposing dynamics of a context in order to effortlessly act within it, both shaped and shaping. Lao Tzu (Chapter 37) states that 'Dao never acts but nothing is left undone; should lords and princes be able to hold fast to it, the myriad creatures will be transformed of their own accord' (Lao Tzu, 1963: 42). In this formulation, nature will always find a way if left to its own devices, and there is no need for interference and control (Chen and Schonfeld, 2011: 199). Nature has its own self-correcting mechanisms.

Wuwei is less risky, with fewer side effects, because it relies on the unity and transformation of pairs of contradictions. Action that arises from only one side of an opposition (for example, yang) can lead to change and reversal as too much action results in a loss of control and imbalance. In the UK, the decade-long imposition of economic austerity contributed to a lack of preparedness for an unanticipated health crisis in the face of a pandemic. Policymaking was polarized around conflicting and contradictory policies relating to the economy and health. Politicans failed to see the extent to which the two were mutually implicated. The either/or approach, particularly in the US and Europe, contributed to a fluctuation of opening and lockdown, the emergence of a second (and in some places a third) wave, against the backdrop of winter, which exacerbated both crises. Weary populations faced degrees of lockdown in much different circumstances. People drove more rather than taking public transportation, which was a step backward for the environment. Isolation was exacerbated by increased restrictions combined with the greater difficulty of spending time outdoors, which exacerbated mental health problems.

The positive purpose of wuwei is to support and allow all things to reach their natural development, which is contrary to reckless and aggressive action. Wuwei pursues the most efficacious action that leads to the best result. Yielding to the conversation, combined with skilfulness, produces a natural equilibrium, which results in an unfolding that involves fewer risks and side effects. Wuwei is less about passivity or not acting than avoiding serious mistakes and disasters. It is more about allowing the desired potentials of a context to unfold effortlessly. The concept provides insight for rethinking how the various potentials of both climate change and the pandemic, which is a focus of the next snapshot, might be successfully navigated.

SNAPSHOT 4

War/No War

Western strategic thought has been heavily influenced by the linear strategy of, for instance, Carl von Clausewitz (Corneli, 1987: 429). Sun Tzu, by contrast, has a distinctly non-linear approach to strategy, grounded in an ancient Chinese concept of time in which past, present and future are simultaneously present in the strategist's mind (Parquettes, 1991: 46). According to Corneli (1987), Sun Tzu's strategy is conceived over a long period of time and constantly revised, engagements are quick when they cannot be avoided and advantage is gained through surprise, flexibility, adaptability and foreknowledge, which are critical given the emphasis on strategy over tactics. Clausewitz instead favoured tactics over strategy, saw surprise as a hindrance and placed limited emphasis on flexibility and foreknowledge. His objective was to apply overwhelming force at the centre of gravity with the intention to destroy the enemy. Sun Tzu's focus, by contrast, was the psychological relationship and the attempt to outwit the enemy. In the first instance he sought to attack the enemy's strategy, followed by his alliances and only as a last resort an enemy's army.

The two strategic traditions are underpinned by different assumptions about the nature of 'reality'. Moving from the linear to the dialectical, the question becomes what it means to work strategically in a situation where the two sides of a conflict, or the two sides of a contradiction, are interdependent and interpenetrated. The objective of destroying an enemy is cast in a different light once the enemy is understood to be bound up with the self. The latter suggests that any victory would be bittersweet. As stated in Chapter 31 of the *Daodejing* (Ames and Hall, 2003: 125):

> When the casualties are high,
> Inspect the battleground with grief and remorse;
> When the war is won,
> Treat it as you would a funeral.

But what does it mean to say that the enemy is bound up in the self? L.H.M. Ling (2014: 120) rearticulated the Buddhist concept of 'inter-being' in Daoist terms. Inter-being, she states, 'shows us how to *be* water, where one drop cannot be separated from another.' Water, like us, she goes on to say, can be both a great healer and a source of destruction. Inter-being leans into the former rather than the latter. At stake is how the relationship between oppositions is understood, that is, whether they are separate and mutually exclusive or entangled and continuously changing in relation to each other. The other, in an entangled relationship, has a different value than an other who is ontologically separate.

The previous snapshot zoomed in on the dynamics of actionless action (wuwei). This snapshot shifts the apparatus slightly to observe the significance of yinyang and the non-presence/presence relationship for strategic engagement in a context of war, as well as Sun Tzu's objective to win without actually fighting a war. Much has been written about Sun Tzu in the West, but rarely with attention to the Daoist roots of his strategy or its dialectical nature. Derek Yuen (2014) is an exception. He provides an in-depth analysis of the various dimensions of Sun Tzu's strategy, with particular emphasis on its Daoist roots and its relationship to different strategic traditions in China. My objectives in this snapshot are much more limited. While seeking to take the Daoist underpinnings seriously, actionless action (wuwei), the focus of the previous snapshot, is brought to a question of yinyang strategy and its implications for navigating a relationship of negative difference, that is, war. A further objective is to bring yinyang strategy to the analysis of a particular contemporary context.

The first section explores the significance of Sun Tzu's framing of an 'art' of war. The second unpacks the strategic implications of following patterns within nature. The third elaborates the strategic dynamics of the *Art of War* as a backdrop for analyzing a contemporary context, that is, the war against COVID-19. The exploration of a non-military context of war highlights the broader applicability of Sun Tzu's strategy beyond military engagement and for the contemporary world. Further, the contours of this war provide insight into the importance of nature and the environment in Sun Tzu's strategy, as well as further perspective on the relationship between the present and non-present, as discussed in the previous snapshot. The enemy in this context is a formless and unseen virus, which is precisely what makes it so formidable. Beyond the objective of victory against the virus, the dialectical dynamics of the context reveal a creative resonance, which points toward different future potentials rather than assuming a return to 'normal'. The negative and positive are bound up with each other in this situation, raising a question of how to effortlessly navigate the long-term strategic potentials that emerge from a context of global pandemic.

Strategy as artscience

Daoism is most often associated with the *Daodejing.* Sun Tzu belongs to a longer tradition of Daoist military strategy. The *Daodejing* is more often assumed to have influenced the *Art of War* than the other way around. However, recent evidence suggests that Sun Tzu's thesis was published prior to the *Daodejing* (Li, 2000; Ho, 2002; Yuen, 2014). The two arose from roughly the same context. The *Art of War* is said to have been written in the late spring and autumn period before 403 BCE (Yuen, 2014: 41). The *Daodejing* came into being during the Warring States period, which began in 403 BCE (Ames and Hall, 2003: 1). In the circular and somewhat paradoxical relationship that emerged, the more practical guide to strategy and action in the world provided substance for exploring in the *Daodejing*, a more expanded view of the world and universe.

While the *Daodejing* is viewed as a philosophical work in the West, Yuen (2014: 68) analyzes it as a strategic text, which is more common in China. The close relationship between the two works, and their shared emphasis on strategy, suggests the extent to which a particular conception of reality, emerging from polarity, is predisposed not only to the strategic (Jullien, 1995: 189) but to 'grand' strategic thinking. Grand strategy, in this sense, resides within an organic holism and involves a systemic approach which explores all positive means and forms of power rather than equating strategy with military strategy, as is so often the case in Western strategic thought (Yuen, 2014: 14). In the context of the late Spring and Autumn period, and the early Warring States period, the most widely circulated texts were military handbooks rather than philosophical or religious texts (Ames and Hall, 2003: 1). It is thus perhaps not surprising that a landmark work on the practice of war would go hand in hand with cultural currents that reinforced dialectical and correlative thinking. Dialectical and correlative thinking resonates with quantum complementarity.

It is interesting that Sun Tzu named his book the art of war rather than the science of war. Art often adopts war as its subject matter, which has been a focus of scholarly examination (for example in Sylvester, 2005) as part of an 'aesthetic turn' (Bleiker, 2009: 271–2) within international relations. Chinese scholars such as Wang Yiwei (2007) have suggested that international relations should be studied as both science and art. He questions the typical analysis of state interactions within a vertical (domestic politics)/horizontal (international anarchy) dimension and an either/or logic. Wang instead suggests that a three-dimensional model would make it possible to conceptualize a more diverse global space, which would account for the Chinese school as well as the Western. Of interest here is his presentation of the three dimensions in terms of space, time and life force, which correspond to science, art and will.

The three-way dynamic is useful for rethinking the 'art' in Sun Tzu's title, as 'artscience'. For any number of reasons, the second translation, by Lionel Giles, which for a half century became the definitive English translation (Luo and Zhang, 2018), may have avoided an English title that referred to both art and science.[1] But, as Wang's argument suggests, there may be a justification in the original Chinese.[2] In the original Chinese title, *Sun Tzi Bing Fa*, 'bing fa' (兵法) refers to 'military law', building on the Chinese character for law (法). But fa can also be interpreted more broadly as theory or principle, as reflected in its use in other fields related to ethics, politics and astronomy (see, for example, Hansen, 1994). In this respect, a strict contrast between art and science is problematic and might be replaced by an artscience polarity.

Roger Ames (1986: 317) argues that Daoism, which proceeds from art rather science, produces an '*ars contextualis:* generalizations drawn from human experience in the most basic processes of making a person, making a community and making a world.' That Daoism begins with art suggests that the yin-like characteristics of art give rise to and shape action within a particular kind of law-like universe. Ames (1986: 320–1) sketches art and science as binary oppositions, as conventionally understood in the West, while also transforming the relationship through their contrast. Science, he argues, emphasizes abstract logical construction and pre-assigned patterns of relatedness, which only register the particular in so far as it fits the pattern. The aesthetic order, by contrast, prioritizes the uniqueness of the particular, concrete detail relating to immediate existence, from which one moves toward generalities within an emergent complex pattern of relatedness that is novel. Different ingredients mix to achieve harmony, as in the culinary example from the previous snapshot.

Logical order is consistent with a Newtonian understanding of the universe as a mechanism. However, quantum mechanics and Daoism bring the polar oppositions into a dynamic relational ordering, which is both art and

---

1 Ancient documents, such as Sun Tzu's *Art of War* and the *Daodejing*, were written in classical Chinese. The introduction of these texts in the West involved a process of double translation. The classical Chinese had to be interpreted, primarily on the basis of annotations and translations, into modern Chinese before translation into English. Given this process of double translation, the potential for mistranslation also doubled. The ancient documents themselves, and more specifically the traditional Chinese characters, are the baseline. I would like to thank Yang Yuanfuyi for this clarification, and for providing Chinese characters and their phonetic equivalents.

2 Notes on translation: As in the last snapshot, I have in places added Chinese characters to enhance the meaning of concepts. Please see endnote 7 in Snapshot 3 for clarification.

some notion of natural or conventional law (science), where the opposites are mutually implicated. Contextual detail drawn from human experience precedes generalization, and expresses a particular kind of natural law that is manifest in the unique and specific rather than the abstract. The pattern arises not from the materiality of things but rather from the dynamic movement of yin and yang. Erwin Schrodinger coined the term *Verschrankung,* or entanglement, which in the German use suggests an unfolding or crossing over in an orderly manner, akin to a finely woven tapestry (Clegg, 2006: 3). The tapestry metaphor fits with an image of the cosmos as an emergent composition which both constitutes and is constituted by the polarity of wave and particle, yin and yang, and the emergence of differentiation and the unique (de) from the whole. The threads are both present in the seen and non-present in their hiddenness beneath the surface of the woven fabric.

Within Daoism, the vast entanglement, and the dynamism of yinyang within it, is called the 'net' or 'warp-and-woof' (*gangji*, 綱紀)[3] of things (Wang, 2012: 50), which in the Chinese character has other meanings that extend to 'social law and order' or governing principle. Like Schrodinger's entanglement, the term gangji comes from an image of fabric, in this case made of silk. *Gang* provides the strand from which other threads are attached, while *ji* represents the mesh of the others. Together they reveal the interrelatedness of the ten thousand things within the net (Wang, 2012: 41). Gangji is the basis for ganying or correlative resonance, as presented in the previous snapshot, which pulsates through the force field of qi that infuses the cosmos. An effective strategy seeks to resonate with the hidden forces at work in any given situation, and to use these resonances skilfully to bring about results (Wang, 2012: 157).

The point is not that art is lacking in form. Rather, particularity or uniqueness is the emergent condition of form that is in a process of becoming, as distinct from existing a priori as matter. Yin and yang are mutually dependent, but yin leads as the background potential collapses into form. From this perspective, science and art are not mutually exclusive, as in a Newtonian binary, but are correlatively related. Strategy becomes artscience and a source co-creativity, from which the familiar dichotomy between world and transcendent can be recast, which then further changes the relationship between science and religion.

3 纲纪.

Destruction or deference

Religion is said to be derived from the Latin word *religare*, which means 'to bind'. Secular refers to the world or worldly. In Western thought, the relationship between the two is one of contradictory opposites. However, once put together in terms of the whole, much like yinyang or artscience, the context changes. Religaresecular may sound nonsensical, but it nonetheless conveys the point that the world and universe, and humans within it, are not and cannot ever exist separately as essences. The world is part of and bound up in the universe just as humans are bound up in the world. The relationship to the universe is not transcendent but immanent. When Sun Tzu refers to heaven (*Tian*, 天) as a component part of the organization of strategy, he does not mean a place outside the world occupied by a god who created it. Rather, in classical Chinese, Tian is the world itself, although the classical Chinese character can also refer to forms of sacred power, an emperor or nature that is infused with transcendent qualities. According to *Zhuangzi*, transcendence is not a state in which we are cut off from the world but rather a realization of our connectedness and interconnectedness within the world (Knightly, 2013: 126). As Ames and Hall (2003: 65) state, '*Tian* is both *what* our world is and *how* it is. Tian is *natura naturans*: "nature naturing"'. All that exists in the world in its diversity is engaged in creating the world. Tian is the emergent order that is negotiated between the many beings, each in their uniqueness. The question that arises is less what heaven (Tian) is than what kind of relationship between humans and their environment is most productive (Ames and Hall, 2003: 63).

We are made through difference and in relation to others. The process of co-creation involves a communication of deference rather than a reliance on reference (Hall, 2001: 247). The deference of the strategic conversation arises from the complex interaction of the opposing forces of yin and yang, which provides insight into the detail of the overall composition. Knowledge of the landscape, of self and other, of the rhythms of nature within it and of planning, are crucial. Being able to see the larger terrain, including the unseen dimensions that lurk in the periphery or are otherwise hidden, as well as the rhythms and resonances that pulse through it, is essential to identifying the potentials of a context such that one might be carried along with it as the situation unfolds. The strategist is both weaver and woven, moving within a tapestry from which she cannot be separated, blending multiple factors into a coherent whole. Humans have ontological parity with all in nature and their thought is no less patterned on it. The strategist seeks power and efficacy that come from harmonizing with the Dao in order to respond and be co-constitutive with the diversity.

The deferential aspects of conversation are easily grasped. However, a relationship of deference is somewhat harder to square with the relational dynamics of strategy that involves complex negative differentiation, that is, war. War involves, to varying degree, destruction of the other. A relationality of deference in war would seem no less contradictory than the idea of winning a war without war. The good strategist embraces and makes use of contradiction and paradox. Deference involves acknowledgement of both sides of the contradiction, and brings both sides of the dynamic into awareness – yin, or the non-present, in particular (Wang, 2012: 145). By doing so, the whole, as seen from one's own position within a broad and complex configuration, comes into focus. All correlative pairs, including those of offense and defence, will be constantly shifting from one end to another (Yuen, 2014: 27).

Deference, as such, while usually associated with a relationship of respect, also has a broader connotation of accounting for or acknowledging the presence and role of someone or something within an entangled relationship. Deference points to the systemic nature of a relationship, within which one engages with other parts of that system, rather than some notion of 'reality' independent of it. As Yuen (2014: 37) notes, 'Subjugating the enemy without fighting' is the best strategy because it causes the least disturbance to the system. The ultimate goal of grand strategy is to restore the system, meaning all of the related parts which compose the whole, to a relatively stable condition so that the fruits of victory can be enjoyed. Even if you don't ultimately like your opponent in battle, your fates are not entirely separate. The good strategist will attach a large degree of importance to the interdependence and interpenetration of two sides in an opposing pair, while attaching even greater importance to the ways in which they can be used to advantage (Li, 2000: 83–4). The composition is not one dimensional but plural and holographic, involving the protagonists as well as the contradictory patterns that arise from nature.

The analysis that follows assumes that the practice of war is internal to or situated within the environment. Both protagonists are embedded within a system. The environment and war, here explored as a dialectical relationship, are usually treated as two distinct topics. The literature on environmental security, as well as the discourse surrounding related security policy, views environmental deterioration as a source of conflict (see, for example, Barnett, 2000; Deligiannis, 2012). A second area explores the impact of war on the environment (for instance, Mannion, 2003; Westing, 2012). In both streams, the environment and war are treated in separation and as being in a causal relationship, that is, environmental change causes conflict, or war causes damage to the environment, respectively. The question of how war damages the environment constitutes the practice of war as external to and impacting on the environment. There has been far less attention paid to the relationship between war as a set of practices and the environment as a context within which these practices are enacted. The latter highlights contradiction and emergence rather than causality.

This different angle rests on a distinction between war and warfare, which is important for understanding the application of Sun Tzu's strategy. Wars may be fought over any number of problems, from the coronavirus to drugs or even the environment itself, but these do not necessarily involve warfare, that is, military deployment.[4] Reference to these phenomena as war is more than just metaphor. Metaphor, like language more generally, gives form and meaning to an otherwise formless phenomenon (Lakoff and Johnson, 1980; Fierke, 1998). War, as distinct from the broader concepts of conflict or violence, is a human artefact based on shared rules, whether these be the constitutive rules of language or the regulative rules of political bodies. From this perspective, as Arundhati Roy (2020) has noted, the war on the coronavirus *is* an actual war, and is treated as such by elites. Her logic might be extended further to explore the regulative rules, such as political directives to stay home in order to avoid contracting the virus and so forth, which were constitutive of the war, as well as the rule-based language and classificatory scheme by which the degree of danger was communicated and understood to be an existential threat, not only by the individual but by the community or even the planet as a whole. The debates about personal protective equipment (PPE) for frontline doctors, nurses and other care providers were not qualitatively different than earlier debates regarding the insufficient supply of protective equipment available to soldiers in Iraq. In both cases, the frontline risked their lives in the protection of others. There is a danger that use of a language of war may contribute to the militarization of a health problem or reinforce the self-interested behaviour of states. War is a human artefact for managing a conflictual entanglement, but it also potentially transforms those entanglements.

War, as a rule-based human artefact, can be distinguished from a different metaphor of water, which expresses the formlessness and rhythms of the natural environment. A water metaphor runs throughout both the *Daodejing* and the *Art of War*. Yuen (2014: 81) states that the water metaphor is the most condensed strategic theory, and applies to all situations due to its simplicity, naturalism and production of patterns that are never definite. If linear approaches to strategy suffer from a lack of flexibility, water never risks being pinned down as it moves; it has multiple potentials rather than a single objective. It is masterful in revealing the pattern of an enemy while concealing its own (Yuen, 2014: 81). If the rules of war provide form to a human social and political activity, the strategic nature of water arises from its formlessness, motion and adaptability. The formlessness of water

4 Yuen (2014: 15) makes this distinction.

adapts to whatever container it is placed in. Water can be both weak and yielding (yin) or strong and forceful (yang). As Ames (1993: 120) states, 'The velocity of cascading water can send boulders bobbing about', which is its strategic advantage.

The human artefact and nature combine in these two images of war and water, the one a product of culture, or more precisely cultures in conflict, and the other referring to a pattern that is universal. For Sun Tzu, the primary objective of war is to avoid war. To seek resolution of the contradiction is to miss a fundamental point of Daoist strategy, also expressed by quantum complementarity, that the paradox or contradiction is itself the dynamic of life and cannot – permanently, in any case – be undone.

The paradox or contradiction must instead be negotiated and navigated. In this respect, the yinyang dialectic, while sharing a family resemblance with the Hegelian, is distinct. The Hegelian-Marxist dialectic, as Ling (2014: 41–2) notes, contains a form of absolutism which both progresses linearly toward an ideal and requires an antithesis that is other, a category within which Hegel lumped all women and 'Orientals'. If revolution is the outcome of the thesis/anti-thesis dialectic of Hegelianism, Daoist dialectics seeks first balance and harmony. As a natural order of change operates from within as well as out, action is non-coercive (wuwei*)* and change progresses organically and contextually, without requiring revolution or violence. Yinyang strategy uses contradiction to analyze the relationship between objects and events, to transcend as well as integrate apparent oppositions and to embrace conflicting viewpoints. The strategy expresses a polarity which eliminates the problem of externality in strategic thought (for example, an 'us' distinguished from external enemies), given that the enemy and situation are understood to be part of an overall system. According to Yuen (2014: 16), the law of non-contradiction, that a proposition cannot be both truth and false, simply does not apply to Chinese thought.

Bound up in the universe and nature, the strategist, on the one hand, works with the dialectic as it arises in any given circumstance, and on the other, seeks to achieve harmony within difference, including in the relationship between individual and world by which the two coexist and become embodied in efficacious human action. The human way (*rendao* 人道*)* and the heavenly way *(tiando* 天道*)* are interpenetrated. Both together follow the pattern of nature, which is the source of all existence. In its original use, Dao referred to a road or path upon which one walks, which extends to a way of acting and living, but which also emphasizes that to follow a path is different from being lost or wandering aimlessly. It is to walk with direction and mindfulness (Wang, 2012: 45). It is to be strategic. Being strategic is not about sheer force and the importance of technology. It is about brain power and outwitting the enemy rather than overwhelming them (Yuen, 2014: 14). While Sun Tzu's strategy was developed in a context of warfare, the emphasis on strategy gives it a much broader application to, among other things, corporate and sport culture (see, for example, Chen, 1994; McNally, 2012; Li-Hua, 2014).[5]

The strategist as artscientist is a part of the tapestry she weaves. She engages from the position of her uniqueness within the pattern of the aesthetic composition, which she must navigate and co-create with others. The strategist navigator is infused with the same life force (qi) that permeates all of nature. The *Huainanzi*, which is a collection of essays based on a series of scholarly debates held at the court of Liu An, prince of Huainan during the Han dynasty,[6] compares yinyang to yu, the skill of steering a chariot and the art of navigating a path for a horse-drawn chariot. According to Wang (2012: 154–5), the skill can extend to navigating any path, but given the dialectic, the process particularly regards the navigation of the relationship between the seen and unseen, which entails learning to easily and artfully locate oneself in relation to one's milieu. Skill at charioteering is less about courage (*yong* 勇) than demonstrating a kind of intelligence (*zhi* 智) and a strategy for becoming an embodied navigator (Wang, 2012: 154). Resonating with the hidden forces at work in any situation means using such resonances to skilfully bring about results. The latter requires attention to the periphery and what is not seen, as well as to that which is present and seen.

Sun Tzu's strategy

The parallel between Daoism and quantum physics provides a physical basis for understanding the dynamics of Sun Tzu's grand strategy by placing it in a world of indeterminism, a life force (qi), entanglement and a relational ontology whereby oppositions are complementary and mutually inclusive rather than mutually exclusive. Quantum science provides a framework for understanding the significance of the Daoist conception of the universe as being more than myth or specific to a culture. While the cultural mindset may be more at home in Asia, we all live in a quantum world. The dynamic relationship between opposites, whether particle and wave or yin and yang, is fundamental to Sun Tzu's *Art of War*. Like quantum physics, yinyang strategy assumes a universe of impermanence and change as well as the radical uncertainty that comes with it. Strategy is about navigating and negotiating the uncertainty (Wang, 2012: 55). The

5 The corporate application of Sun Tzu's thought provides an interesting contrast to the liberal rational economic man who is bent on utility maximization and without emotion. The feeling-thinking dialectic within Daoism suggests a place for emotions in Daoist strategy.

6 The *Huainanzi* explored not only Daoist concepts but Confucian and legalist concepts as well, and, among others, theories of yinyang.

navigation process relies on intelligence that is larger than the individual, within an organic self-generating system within which universe, world and nature are interpenetrated and in a resonant relationship (Wang, 2012: 46). The presence of the Way (Dao) isn't metaphysical but flows from the qi that infuses mind/body. Sun Tzu, who is considered to be one of the world's greatest strategists, provided a practical guide for navigating the universe and the rhythms of nature, with the objective of outwitting the enemy and thereby avoiding military engagement if at all possible.

Sun Tzu's strategy is heavily dependent on planning and knowledge of the context of potential action. Qi, the life force, and the mind, though intangible, 'constitute the "information" or "intelligence" that Sun Tzu considers most important' (Yuen, 2008: 190). Consciousness relates in part to what today is referred to in security circles as intelligence. Given Sun Tzu's reliance on a, roughly translated, 'divine skein' (Warner, 2006: 486), intelligence is arguably more than just information about the other, although the concept does relate to spying. Sun Tzu elaborates five types of spy, who work simultaneously, with none of the separate parts knowing the overall method of operation. Intelligence for Sun Tzu is first and foremost about knowing the context. The point is to identify patterns of nature and the potentials within them in order to effortlessly and successfully achieve one's objectives. Similar to gangji or Vershrankung, a skein is a length of yarn or silken thread coiled loosely or in some sort of reel. Sun Tzu's reference to divine spirit suggests the value of the web of intelligence as a 'treasure of the ruler' and a system by which she coordinates, secures and exploits the findings of the agents (Warner, 2006: 486). Qi, as life force, relates to a form of intelligence that pervades all life. Sun Tzu's emphasis on knowing not only the enemy but the environmental conditions, for example the terrain, points to the importance of this broader context of intelligence. Knowing the enemy while remaining 'formless' and adaptable oneself makes it possible to attack when the enemy is at its weakest point, throw them off balance and thus achieve victory with minimal effort.

The following section elaborates a set of categories and a framework that relies on those aspects of Sun Tzu's strategy that are most relevant to its application to the war on COVID-19. By way of an interface with the other snapshots, it is worth mentioning the role of meditation in Daoist military strategy. Yinyang strategy involves careful analysis of the potentials of a context in order to navigate them with calmness of mind, honed through practices of meditation, even in the midst of a storm. Techniques of mind, not least breathing meditation, provide a way for the wise strategist to delve into her own nature, finding calm within the constant stream of thoughts, feelings and perceptions that normally condition consciousness (Meyer, 2012: 60). Consciousness that merges with the Way (Dao) is free of bias, misconception and delusion. The practitioner is able to see clearly, to walk straight ahead with purpose and direction. The strategist, as a 'genuine person', is committed to practices that seek to penetrate through to the Way of their own person (Meyer, 2012: 61). The 'root and branch' metaphor formed the heart of ancient military strategy. The Way is the root, and all contingent phenomena are branches which are replicated at every level of existence, including matter, energy, the mind, the body, human history, society, culture and politics.

The rhythms of nature

The rhythms of nature flow from the root to the branches. Foo Check Teck (1997), in his reconstruction of Sun Tzu's reminiscences, points to numerous examples from nature that may have informed practices of warfare. The hard shell of the tortoise, for instance, provides a model for constructing villages with protective walls, Just as the pangolin, when threatened with a stick, rolls into a ball of scaly armour, the soldier responds to immediate threat with a spray of arrows or a metal shield. Insects are masters of the art of camouflage, changing the colour of their skin to harmonize with the bark of a tree they land on or spinning fine webs, almost invisible to the eye, in order to trap their prey. As the spider only fights or entraps the insect when it is completely exhausted from fruitless struggle, the general only sends his troops into battle when the enemy is demoralized and tired. Caterpillars reveal a further aspect of adaptive change in their transformation into a chrysalis from which, days later, they emerge as a butterfly. The caterpillar changes into something radically different, just as the farmer, through training, is transformed into a mighty warrior. Spiders practice the art of deception in laying traps that capture their prey, much as humans use nets to catch fish. In the rhythms of nature, Sun Tzu finds forms of protection, deception and change.

Water, the essence of strategic theory, is also part of the rhythm of nature, and provides insight into the relationship between form and formlessness. Sun Tzu (Sawyer 1994: 193) develops the significance of the water metaphor for understanding a strong army's ability to change form:

> The army's disposition of force (*hsing,* 形)[7] is like water … Water configures (*hsing*), it flows in accord with the terrain; the army controls its victory in accord with the enemy. Thus the army does not maintain any constant strategic configuration of power (*shih*, 勢),[8] water has no constant shape (*hsing*).

[7] Yuen uses the Wade-Giles romanization. The Chinese Pinyin is *xing*.

[8] 势, the Chinese Pinyin is *shi*.

Sun Tzu (Sawyer 1994: 193) further states that if the pinnacle of military deployment (*hsing*) approaches the formless, then 'even the deepest spy cannot discern it or the wise make plans against it'. *Shih* and *hsing* relate to other oppositions such as emptiness (*hsu*, 虛)[9] and solidness (*shih*, 實),[10] which involve creating the illusion that no position or force is permanently empty or solid, thereby presenting the opponent with an ongoing dilemma of what to defend or what to attack.

The reversal of opposites

The yinyang dynamic exists in nature. The strategist seeks to flow with the rhythms of nature. Far from suggesting passivity or determinism, the strategist, with objectives and the terrain clearly in mind, effortlessly negotiates and navigates a context. This navigation works with the dynamic interplay of opposites rather than against them. The strategy might involve both the more direct and conventional, that is, that which can easily be seen, as well as more unorthodox strategy which arises from working with the unseen. Concepts in the *Daodejing* from the natural world – that things, while remaining in a complementary dynamic tension, revert to their opposite after reaching their extreme – resonate with Sun Tzu's dynamic characterization of the unorthodox turning into the orthodox (Sawyer, 2007: 55). The mutual implication of opposites, and their potential reversal, is key to one of the *Art of War*'s most famous passages, that 'warfare is the Dao of Deception' (Sawyer, 1993: 158). Sun Tzu instructs the reader, 'Although capable, display incapability. When committed to employing your forces, feign inactivity. When your objective is nearby, make it appear as if distant; when far away, create the illusion of being nearby' (Sawyer, 1993: 158). The 'Dao of Deception' provides a larger framework for understanding the string of oppositions contained within Daoist military strategy, all of which relate to fundamental Chinese philosophical concepts.

The Dao of deception rests in particular on the manipulation of *ch'i* and *cheng*. Ch'i translates as 'crafty' or 'unorthodox'. The unorthodox or crafty (ch'i, 奇) refers to the unusual, unexpected, marvellous or strange, in contrast with the more orthodox or straightforward (*cheng,* 正)[11], or more simply, a contrast between indirect and direct strategy. Sawyer (2007: 5) recounts Tien Tan's multi-state, unorthodox strategy to undermine the enemy's will while

9 The Chinese pinyin is *xu*.

10 实, the Chinese Pinyin is *shi*.

11 I have used the Wade-Giles ch'i/cheng rather than the Chinese pinyin qi/zheng in relation to indirect and direct strategy as this makes it easier for the English speaking reader to differentiate qi as the lifeforce (氣/气) from qi as indirect strategy (奇).

rebuilding the spirit of the defence whereby only seven thousand exhausted soldiers and another ten thousand inhabitants trapped in Chi-mo defied a siege of some one hundred thousand. At the heart of indirect strategy is a question of how to exploit an enemy's vulnerabilities (Sawyer, 2007: 378), for example through the manipulation of what is seen. As mutually inclusive oppositions, cheng can become a ch'i force and ch'i can become a cheng force. Ch'i and cheng relate, among other things, to the ability of the strategist to concentrate his own forces while dividing those of the enemy. It is here that the manipulation of seen and unseen is key, as the general, for instance, creates the illusion of attacking in one or multiple places, thereby dividing enemy forces, while concentrating his own at another location.

Leadership and planning

As a military strategist, Sun Tzu was perhaps less concerned with morals per se than with the efficient use of resources or effectiveness in battle. Having said this, moral law is one of five fundamental factors of Sun Tzu's strategy. By moral law he means that which brings people into alignment with their ruler so that they will follow her without fear, although this law also suggests its opposite, that is, that an ability to undermine the coherence of popular will can debilitate an enemy, as was, for instance, evident in Soviet efforts to destabilize and demoralize Western societies during the Cold War (Corneli, 1987: 441). The further factors include 'heaven' (the working of natural forces on the conduct of military operations), the earth (working with the terrain rather than against it), command (the virtues of the general) and doctrine (the organization of the army), all of which highlight the importance of planning (Khoo, 1992: 2–3). In laying out the five fundamentals, Sun Tzu claims that the success or defeat of either side will depend on whether they accept or reject his advice.

The holistic approach to planning places emphasis on creating the conditions by which an outcome can come about on its own, effortlessly. The strategy tunes into the dynamics of a particular context rather than creating a blueprint that may be unable to adapt to a continuously changing environment. For Sun Tzu, at the start of battle one should 'be as coy as a virgin; when your enemy lowers his guard and offers an opening, rush in like a hare out of its cage and the enemy will be unable to defend in time' (Khoo, 1992: 45). The timing of these manoeuvres, as well as attention to weather and terrain, is crucial. Skills in knowing nature's timing and human affairs, as well as knowing the features of terrain, are among the most important possessed by a general (Cleary, 1989: 42). Sun Tzu might suggest, for instance, subjecting the enemy to the worst weather conditions, making their troops march out in the middle of a storm, while one's own troops are comfortably sheltered behind city walls (Foo, 1997: 157). In the meantime, the general's own troops await better climatic conditions that will support their actions. The master strategist knows how to wield the weather and terrain as a weapon, using an onslaught of weather events to weaken the enemy, followed directly by a further onslaught, without warning, while the enemy has its back toward a deep gorge of flushing waters (Foo, 1997: 160). The ability to manipulate the orthodox and unorthodox is key to success in a context where one is severely outnumbered.

The emotional dynamics of war

Balancing the relationship between the qi or ch'i of one's own force and that of the enemy is also key. The other meaning of ch'i as life force is expressed in a single passage in the *Art of War*, and is the earliest reference to be found in military manuals: 'The *ch'i* of the Three Armies can be snatched away; the commander's mind can be seized' (Sawyer, 2007: 170). During the classic Warring States period, military writings developed a martial motivational science centred on the concept of ch'i (氣)[12] that articulated numerous measures and methods for moderating the army's energy and controlling the soldier's commitment (Sawyer, 2007: 50). Ch'i was at the heart of a psychology of fear and courage which involved going to battle at the point that the ch'i of one's own forces was strong and that of the enemy was weak (Sawyer, 2007: 49). The strategist is engaged in a dynamic process whereby she must first and foremost be concerned with conserving her own ch'i while also trying to identify the psychological, moral and political qualities that make up the adversary's ch'i in order to dissipate it or attack at the most propitious moment (O'Dowd and Waldron, 1991: 29), when the energy of the enemy has abated. The wise general seeks to instil doubt and weakness in the enemy while exercising control of the spirit of one's own army. Wei Liao, a strategist in the fourth century BC, believed that the army's ch'i essentially determined a battle's outcome (Sawyer, 1993: 229, 236).

After the victory, when one's own forces are exhausted, Sun Tzu emphasizes that it is far better to rebuild or restore what has been gained than to destroy it (Khoo, 1992: 48). There are several possible reasons for this emphasis. The first regards the dangers of protracted war. Coker (2003: 18) argues that Sun Tzu first and foremost feared the consequences of escalation in war, as this would potentially end in total defeat rather than total victory. After victory, an exhausted army can subsequently be attacked by a third party or be unable to police the peace. Sun Tzu was very clear that no country has benefited from protracted warfare.

12 气.

The second is a moral position, consistent with a maxim, repeated later in texts of the Warring States period and early Han dynasty, of 'sustaining the perishing, reviving the extinct', which became a litmus test for the legitimate use of coercive force and found expression in the idea that victorious rulers shouldn't profit from the resources of a vanquished state but should return the assets to the people in exchange for submission; a claim that the annihilation of a state and its sacred institutions through terror and violence defies deep structural patterns of the cosmos, or that a tyrant who launches an aggressive campaign would experience misfortune (Meyer, 2012). Several of these points relate to the correlative resonance of ganying, which was discussed in Snapshot 3.

A third regards the focus of grand strategy on the system and concern about the unintended or undesired consequences of prolonged action beyond what is needed, not least the consequences of the hatred generated by war. Chapter 79 of the *Daodejing* raises a question of how reconciliation can be successful if, in the face of peace between enemies, enmity remains. Yuen (2014: 95) highlights the amount of time it takes to reverse the consequences of hate, which suggests that excessive measures or prolonging conflict will be counterproductive and disrupt the overall harmony of the system. Hate is an emotion that arises in the context of a relationship, and in the argument of Sun Tzu, is best avoided to the end of preserving the system within which the relationship is entangled.

Unlike Confucius or Gandhi, Sun Tzu does not reject violence all together. Indeed, he sees war as a necessary evil, but argues that the most effective strategy will avoid war all together. It is better to subjugate other states without actually engaging in armed combat, through deception or diplomatic coercion, thwarting the enemy's plans and alliances and frustrating their strategy (Sawyer, 2007: 57). The objective of avoiding the use of force if possible is evident in how Sun Tzu devises the strategy of attack, beginning with the enemy's plans, then their alliances, followed by their army. Attacking fortified cities is at the very bottom of the list. As Sun Tzu states, 'Fighting to win one hundred victories in one hundred battles is not the supreme skill. However, to break the enemy's resistance without fighting is the supreme skill' (Khoo, 1992: 9). Placing these claims in the larger context of Daoism reveals that a central objective of grand strategy not only regards the efficient use of resources, given that war is costly, it further points to the primacy of complementarity, by which oppositions (yinyang), including self and enemy/other, are always mutually implicated and thus in a relationship whereby fates cannot be entirely separated, as expressed in Sun Tzu's maxim: 'If you know yourself and know your enemy, in a hundred battles you will never fear the results' (Khoo, 1992: 11). Minimizing recourse to the use of force is bound up in the further objective of restoring balance and harmony to the system.

Navigating yinyang

The COVID-19 pandemic is an interesting site for exploring the yinyang dynamics of Sun Tzu's strategy. First, the enemy in this war, a virus, emerged from nature, raising important questions about the relationship between nature, environment and the virus. Second, the yinyang dialectic reveals the creative resonance of a context that is larger than human life, and which thereby minimizes the latter's presumed ontological primacy. Third, the context unfolded with the reversal of a number of dramatic and contradictory oppositions. Finally, Sun Tzu's strategy highlights the multiplicity of potentials in a context that has, more than recent wars and within a shorter time, involved a massive loss of human life. Sun Tzu's strategy provides insight into how humanity might move more effortlessly toward a different kind of future than would have been thought possible before the pandemic began. The question is the extent to which the strategic potentials of a context of radical uncertainty are seen and successfully navigated. Sun Tzu's recommendations rest on different assumptions than those of scientific modelling. Leaders across the world have stated frequently that their response to the virus was guided by science. How would the response differ if guided by artscience? Sun Tzu starts with the context, rather than the abstract model, and the identification of dialectical patterns that drive the process of differentiation within it. What follows, within space limitations, is an attempt to map the dynamics and draw out some of the strategic implications.

One would not normally think of a virus as an agent.[13] Agency is usually associated with human action. Given a distancing from the concept, in light of its association with the human individual, we will refer to both 'virus actant' and 'human actant', to highlight the impact of their mutual movements on each other. Within Daoism, humans have no ontological primacy. Action and strategy are about negotiating one's environment and acting in accordance with the rhythms of nature. On some level, action and strategy thus belong to all life, down to the brainless brittle star (Barad, 2007: 270). From this perspective, the war is less about humans versus nature than about an imbalance between human artefacts and nature. It arises from a conflict between human strategies that have imposed a particular kind of form on the world and an invisible formless enemy arising from nature. Knowing the enemy is a crucial aspect of Sun Tzu's strategy. We start by looking at the virus and the dynamics of its spread, after which we will move to the human response to the virus, at least in the early stages, and consider why or why not, from the perspective of Sun Tzu, the strategy of the latter was successful.

---

[13] Although the notion of non-human agency is a theme in the post-humanist literature.

Knowing the enemy

Effortless success against an enemy requires knowledge of them, as well as of the self. Knowing the enemy is important for understanding the context, the systemic relationship and thus the parameters of engagement. Knowing the enemy was the biggest challenge in the war against COVID-19. As a new strain of the coronavirus, COVID-19 was not only invisible but unknown. There was no prior experience to draw on in assessing how this particular virus would impact on human populations. The enemy took the offensive, catching populations, and particularly populations in Europe and the US, by surprise.

The enemy was astute because *attuned to the patterns of nature.* COVID-19 is a virus. Thought to have arisen from wild animals, such as bats or pangolins in a wet market in China, the enemy was invisible and formless. It *adapted to the circumstances*, whether flowing freely and unimpeded, like a forceful cascade of water down a hill, or contained and redirected, like the waters of a flood. The virus, in its *formlessness*, adapted to a continuously changing global environment and in response to the diverse strategies adopted by humans to defeat it. The virus *acted effortlessly*, spreading and infecting without detection, making itself known only at a point when the damage had become too significant to be ignored. The virus was both non-present and unseen. It caught its victims unaware, only to become manifest in their symptoms, illness and potential death. Knowing the human enemy's weakness for affection and travel, the virus *used resources efficiently* as it was effortlessly carried to new destinations by its eventual victims, expanding its presence with more or less ease in diverse locations, depending on the more localized response. Populations were like flies walking into an intricate spiderweb, only to find themselves entrapped once it was too late. The virus and the human became entangled as the former invaded bodies.

The enemy was skilled at *working with the Dao of deception.* The wise strategist creates the appearance of preparing for battle in an area that is not the ultimate objective, as a way of diverting and dividing the enemy, such that one's objectives are easily realized elsewhere. The art of camouflage and diverting the enemy's attention is crucial to this. As such, the virus itself was a mere foot soldier. One need only ask who ultimately benefits from the deception. The virus, through its rapid spread, clearly did grow strong and increase its influence, quickly bringing societies and economies to their knees. But nature, from which the virus emerged, was itself the greatest long-term beneficiary.

A multi-front war, involving both direct and indirect strategy, had the effect of exhausting the human enemy, wearing down their ch'i or morale, at which point the virus rushed in like a hare to maximum effect. Weather events were warning shots in this war. Repeated and deadly fires in Australia and California, floods in Europe and Asia, hurricanes of increasing velocity in Asia and the US and the occasional deep freeze could be seen in the months and years prior to the unveiling of the virus. When the human enemy did not listen or respond with deference, nature reverted to the more indirect strategy of unleashing a virus which forced them to listen. These were not the actions of a god-like transcendent figure, or of spirits, but rather a result of correlative resonances arising from the tremendous imbalance between nature and human artefacts and a reversal of opposites by which direct strategy becomes indirect and vice versa. With the emergence of the Anthropocene, the imposition of human form on nature was approaching completion. The environment had been silenced and destroyed, or so it seemed. Drastic action was needed or it would be too late.

Nature took the offensive in a war that it, by all appearances, was losing. Human artefacts, from automobiles to airplanes and industry, were leading not only to the destruction of nature but resulting in significant deaths among the human population due to the impact of, for instance, air pollution among other things. Although weak and very yin-like, the virus quickly brought about an almost global lockdown, which revealed a series of contradictions. Air pollution, a byproduct of the human artefacts, attacks respiratory systems and is particularly toxic in the world's most densely populated cities. Nature unleashed a virus making lungs already damaged by air pollution doubly vulnerable, but at the same time displayed compassion for those victims, suggesting that it was indeed acting ultimately on their behalf rather than against them. As humans were forced into lockdown, carbon levels dropped at an unprecedented rate and the inhabitants of large cities, who in the past may have worn masks to guard against pollution, became able to breathe more freely, albeit they were now forced to wear masks to stave off the virus.

The economy was another front where the indirect offensive of the virus was lethal and led to several reversals from one extreme to another. As highlighted in the previous snapshot, the governing party of the United Kingdom reverted to policies that were contrary to past Conservative practice, paying a significant proportion of the salaries of those who would otherwise have been unemployed due to the lockdown. Here it is useful to reveal a further aspect of the reversal, relating more directly to the imbalance between human artefacts and nature. The Daoist emphasis on ontological parity means human do not have primacy. Humans themselves are a part of nature and entangled with all life. Sun Tzu's prescription to avoid violent engagement unless absolutely necessary mirrors the rhythm of nature. Non-human animals generally don't engage in gratuitous killing.[14] They

[14] Studies have shown that primates, including humans, are the most violent animals (Wanjek, 2016).

are violent when necessary, whether for protection or food. By contrast, the virus revealed the structural violence at the heart of the human economic system, which targeted the weak in particular.

The war started with an awareness that all were one in their vulnerability to the virus. However, as the virus spread more widely, and the deaths mounted, significant inequalities became evident. In the UK and US contexts; for instance, those risking their lives and dying on the front lines were disproportionately of Black, Asian and ethnic minority heritage. As the virus spread, it became clear that frontline workers were not only less valued and underpaid but were more likely to die. Wealthier individuals in white-collar jobs were able to continue working safely from home, and, along with the poor, benefitted from government schemes to address the large-scale unemployment. The brutal murder of an African American man in Minneapolis, George Floyd, at the hands of a White police officer sparked a wave of protest that spread quickly across the United States. Populations, weary from weeks of lockdown, flooded into the streets in protest, revealing the racial fault line and desperation that run through the wealthiest country in the world. The latter comes into focus in Snapshot 6.

The virus managed to achieve what few empires in history have managed, and far more quickly; that is, a global presence and command to which societies kneeled in deference. The global reach of lockdown meant that planes stopped flying, streets were empty of automobiles as people were confined to their homes, the market dropped out of the oil industry, in part due to the absence of demand, and economies went into a tailspin. As a result, the air became clearer. As the virus went on the offensive, animals, and nature more generally, became present and visible and their beauty valued, not least due to the absence of other, more conventional distractions. The rhythms of nature, the waxing and waning of the sea tides, the mating calls of birds at dawn, the kangaroos in the streets of Australia, refocused human attention in forced conditions away from the preoccupations of a highly stressful industrialized life. The stillness and beauty of nature, at least for those fortunate enough to have any nature in their immediate surroundings, opened up a potential for deference to and co-creativity with (rather than destruction of) the environment.

The human assault on nature has involved the destruction of biodiversity, the felling of the rainforests and the accumulation of plastics in the sea and pollutants in the air. The environment fought back. An unseen enemy manifested itself in the presence of human life, skilfully manipulating the relationship between seen and unseen and temporarily bringing a halt to some manifestations of the human artefact. The pandemic was like the imposition of timeout to give human societies an opportunity to stop and, in the emptiness of lockdown, reflect on their own actions. In this respect, COVID-19 put humanity on a state of high alert, a war footing, as it responded, but the ultimate objective of nature was to transform the human enemy into a friend and foster a relationship of mutual deference. While structured around war, the response revealed contradictions and reversals. The three-front attack, involving weather, disease and the economy, exposed the severe imbalance between the human artefact and nature, and resulted in further reversals to opposite extremes.

Knowing the self

It is important to know not only the enemy in war but the self as well. The protagonists in any conflict are fighting with each other, and thus their fates become intertwined. But this begs the question of who exactly the self to be known is. The virus took the offensive, attacking humanity across the globe. In this respect, the virus was both yin – adaptable and formless – and yang – attacking with strength and power. Within a global context, defined by a capitalist global economy, the onslaught exposed more about who 'we' are than could possibly have been seen in its absence, given our tendency to turn away from suffering or to push it from sight. The liberal capitalist self is ultimately the individual who, beginning with a notion of what is 'mine', seeks to maximize its utility. Approached from the perspective of global political organization, the other self in this construction was the nation state, which relies on the distinction between citizens and those who are not full members, such as refugees and migrants. In the decades prior, many states hardened their boundaries to waves of outsiders trying to get in. This was particularly true of states that had been historically most successful in accumulating wealth, not least because of an imperial past, as distinct from, for instance, states in the Middle East, such as Jordan, which took in a disproportionately higher number of refugees fleeing war in Syria.

While human rights legislation contains ideas of global citizenship and belonging, in practice the accumulation of wealth in the hands of individuals with the deregulation of the global economy has led to an extreme imbalance of wealth, with a handful of individuals owning more value than the poorest 50 per cent (Elliot, 2017). The imbalance is as internal as it is external. Following a global banking crisis in 2008, policies of austerity in some of the wealthiest countries had already led to a proliferation of food banks visited by increasingly desperate populations. Populations suffering a sense of loss were susceptible to political propaganda via social media. The propaganda magnified demands to shut down borders to keep refugees and migrants out. Populations had already been sufficiently weakened and divided, not by nature but by existing human structures, practices and artefacts. As the virus swept across the world, suffering was hidden in the dark spaces of lockdown, whether in care homes or hospitals, hostels for the homeless or refugee camps on the edges of the seen. Spouses became more vulnerable to violent partners. Children became a danger as potential carriers of infection to the elderly.

The virus is said to have emerged from a wet market in China. From the first outbreak in Wuhan, China, in 2019, it travelled with human carriers to Iran and Switzerland from which it continued its onward journey. Following an unusually bad winter of weather events, the virus arrived in the West. The impact of the virus was less severe in Asia for a variety of reasons, from the severity of the lockdown in China to more effective testing and tracing systems in South Korea and established norms of wearing facemasks. The UK and US, which eventually competed for top place in the global mortality statistics (per capita), saw a number of reversals of policies to the other extreme. The divisive politics that followed on the Brexit referendum and the election of US president Trump in 2016 gave way, during the first wave of the pandemic, to a new solidarity, which emphasized the importance of societal compassion and a sense that we were all in this together. The key soldiers on the frontline, that is, nurses, doctors and care workers, as well as those delivering food and aid to homes, were celebrated for their courage and compassion in caring for a society in lockdown while risking their own lives.

This section started with the point that the virus was both yin and yang. Its force in causing destruction was visibly present, yet its power arose from its formlessness and ability to spread unseen. The dynamics of yin and yang also ran through the human actant, although for a number of reasons, the analysis is structured around the yin. Yin is fundamental to yinyang strategy. While yin tends to be gendered as female, it is first and foremost about the non-presence/presence relationship and the importance of taking the unseen into account. But use of 'her' also fits with the reversal of the historically gendered roles of the soldier to the frontline figure of the nurse and her role in caring for the sick. The transformation was captured by the British artist Banksy in a painting for the NHS. The painting shows a male child who has dumped his male superhero dolls in the bin and replaced them with a female NHS nurse as the hero that children should admire. Other images in the media communicated a similar message, for instance of male soldiers, now wearing aprons and rubber gloves, delivering PPE to the front line. The images reveal the dramatic reversal from the 'politics of hate and separation' as expressed, for instance, by the politics of Brexit. The front line was made up of doctors, nurses, care workers and all those who served the community in different ways, for example by delivering food and medicine, and expressed a politics of care and compassion.

A stark contrast between two styles of leadership expressed a further potential reversal within this war. Sun Tzu emphasizes the importance of discipline and a strong general to enforce that discipline. At the global level, there was an absence of coordination. Those who would have exercised global leadership in the past, not least the United States, were fuelling division.

The US Trump Administration decided to withdraw from the World Health Organization and engage in a blame game with China, referring to COVID-19 as the 'China virus', thereby turning the global pandemic into an inter-state issue.

The most effective leaders in dealing with the pandemic were women, such as Jacinda Ardern in New Zealand, Tsai Ing-Wen in Taiwan, Angela Merkel in Germany, Erna Solberg in Norway, Nicola Sturgeon in Scotland and Halimah Yacob in Singapore (GNP, 2020), although not necessarily so, as the prime ministers of South Korea (Chung Sye-Kyun)[15], Thailand (Prayut Chan-o-cha), and Vietnam (Nguyen Xuan Phuc) were male. Based on a data set of 194 countries, Garikipati and Kambhampati (2021) found that 'COVID-outcomes were systematically better in countries led by women and, to some extent, this may be explained by the proactive and coordinated policy responses adopted by them.' But the leadership issue was less about men or women per se than gender and the qualities that make for success in a context where the front line was engaged, not with violence against a human enemy, but rather with care for the vulnerable and preventing spread of the virus. Empathy and an ability to listen, hear and to take advice are essential qualities for bringing and keeping populations on board and for inspiring trust. The best strategy empowers public engagement. Foo Check Teck (1997: loc 160)[16] notes that:

> The [Queen] who prevails will be known by the people for [her] strong moral fibre, given that it is in the nature of things that good will prevail over evil in the long run. 'The [Queen] whose words are widely held to be worth their weight in gold is more able to secure compromises than one whose words nobody trusts. The likely winner, in a contest between [Queens], will be the one considered to be most reliable, honourable and trustworthy.'

The notable gendering of effective leadership stood in stark contrast to the continuing role of a hyper-masculine style of politics which had been active in several countries over the previous decade, and which threatened to turn the problem into international politics as usual. In this respect, casting the response to COVID-19 as a war could potentially have smoothed the way towards more conventional notions of war related to military violence, and

[15] Although, it was a female director in chief of the Korean Centre for Disease Control and Prevention, Jeong Eun-kyeong, who has been praised as an 'anti-disease hero' (Walker, 2020).

[16] Foo Check Teck's text uses the 'King' as subject. I have changed this to 'Queen' in line with the gendering of the effective leader in this context as feminine.

was thus better avoided. In addition to identifying the virus with a specific country, that is, China, a vitriolic male leadership, from Trump in the United States to Bolsonaro in Brazil, brought arrogance to war with an enemy that they did not take seriously, all while assuming their own invulnerability. Continuing to shake hands and refusing to wear face masks, practice social distancing or listen to advice, these leaders flouted the very rules that were essential to curbing the virus.

Flouting the rules was particularly problematic given the importance of a trusting relationship between leaders and populations and the need to bring the latter along through trust and communication. When, in Britain, the top advisor to the prime minister flouted the rules and neither resigned nor was fired, a fragile trust with the population was broken. The arrogance of the hyper-masculine leaders made their populations the most vulnerable in the world as was reflected during the second wave in soaring death tolls, reaching three thousand a day in the US, approximately the same number of deaths from the attacks on 11 September 2001. Commentators on the *Art of War*, Zhuge and Liu (Cleary, 1989: 42–3) emphasize that generals should avoid arrogance: 'For if they are arrogant they will become discourteous, and if they are discourteous people will become alienated from them. When people are alienated, they become rebellious.'

The arrogance first took the form of an unwillingness to take the formless enemy seriously; subsequently, their respective publics suffered more. In addition to the larger number of deaths, their populations were weighted down by the psychological and economic impacts of multiple extended lockdowns, precisely because measures were imposed too late and thus continued for longer than would otherwise have been the case, thereby exacerbating the economic impact. Health and economy were treated as an either/or choice, when in fact they were mutually inclusive: a healthy economy requires a healthy workforce; leaving the economy open as cases were escalating meant a more extensive lockdown once it was out of control. For the wise strategist, timing, decisiveness and trust are everything. Qualities that inspire trust were far more evident among leaders of countries that were, relatively speaking, successful in keeping the virus at bay.

The countries that were successful acted early and decisively. They communicated clearly with and to their populations about what they needed to do. The potential success of specific strategies can be analyzed in terms of the flooding example explored in the previous snapshot. A dike will contain a flood of minimal proportions, but the barriers will be breached once the flood becomes too severe. New Zealand, a small island country, went into lockdown almost immediately once the virus was discovered there and very quickly managed to eliminate it. By contrast, the UK, also an island country, albeit larger, left its airports open and delayed lockdown until the virus had taken sufficient hold in society that the health service risked being overwhelmed. While the latter was prevented during the first lockdown, the lateness of the decision added to the deaths not only from COVID-19 but from other illnesses that were not treated during the worst of the pandemic because of the redirection of resources. Germany (at least during the first wave) and South Korea were also successful with a testing and tracing strategy. The latter avoided lockdown all together.

Testing and tracing is more akin to a strategy of redirecting the flood water. It was less a case of redirecting the virus to other places where it could be useful, as in the flooding example, than of identifying individual cases early on so that anyone potentially infected could be isolated and the impact of the virus dissipated without the need for extended lockdown or indeed lockdown at all. In the United States and United Kingdom, widespread testing and tracing was only introduced once the virus already had a foothold in the society, with too many infections for the strategy to be workable. Both countries also opened from lockdown while the virus was still circulating widely, and thus faced an even more virulent second wave. When populations were at their weakest, after close to a year of continuously changing restrictions, vaccines began to be approved, but the virus also mutated and became more virulent in its ability to spread, creating a race against time to vaccinate populations.

The absence of planning and the unwillingness to take the enemy seriously exacerbated the problem. US president Trump had, prior to the pandemic, dismantled the agency that had earlier been set up to deal with pandemics, and the UK had not kept its stockpiles up to date. The US withdrew from the World Health Organization at the height of the pandemic. President Trump turned what was a global problem requiring global coordination and cooperation into a politics of us and them, first directed at other countries, such as China, or immigrants, but ultimately exacerbating racial divisions within the US. With the election of Joe Biden as president, the message went into reverse, from divisiveness to unity in the struggle, but in a context where significant damage, not least due to the polarizing hate, had already been done. The degree of suspicion generated by the early response made for a difficult road back. The hyper-masculine leaders communicated confusing messages to their societies. While claiming to be guided by science, the lack of transparency regarding which science or which model, as well as political messages that contradicted those of medical advisors, meant that the messages communicated to publics were both confused and unclear, which eroded trust and made it difficult for populations to know how to act.

Conclusion

Sun Tzu's emphasis on planning seems contrary to the importance of adaptability but is actually central to it. The wise strategist knows the landscape and uses nature, as well as the weather and the terrain, to maximum advantage. She is also acutely aware of the importance of timing. In order to effectively and effortlessly navigate a context that is constantly changing, in-depth knowledge of the environment and the enemy is necessary. One of the most crucial areas of navigation is emotion, which includes maintaining the morale of soldiers or civilians who are essential to realizing one's objectives while weakening the morale of the enemy. Intelligence is crucial for uncovering those unseen aspects, which may impact on how the situation unfolds, and for thinking about how to navigate the dialectical relationship between seen and unseen in ways that will weaken the enemy, and therefore increase the likelihood of victory without a substantial loss of life. For instance, nursing homes, as a site of close contact and care for a vulnerable population, would be, one might anticipate, unusually susceptible to the virus, and could have been on the radar at a much earlier time. These distinct aspects of planning provide a litmus test for measuring the success of different strategies in tackling the viral opponent.

The war against COVID-19 gave rise to very strong emotions of grief, anger, hatred and revenge, which may have long-term consequences, but also to feelings of compassion and greater awareness of structural inequalities within society and across the world. Winning the hearts and minds of populations is crucial. It is important to avoid creating resentment that may later be turned against you, and to foster gratitude that might work in one's favour. The emotional dynamics also highlight the role of meditation in the Daoist tradition and calming the mind. The wise strategist does not need to repress or deny her own emotions, but she does take some distance from them, to avoid allowing them to cloud her judgement. She instead uses them as her guide. She not only knows herself and understands the importance of acknowledging her own emotions in order to avoid being ruled by them, she also seeks to know her enemy, not least in order to assess their strengths and weaknesses, in order to wisely engage with them as the situation requires, but always with the yinyang non-presence and presence relationship in mind.

The impact of the pandemic on societies across the globe was not uniform, nor was the distribution of the eventual vaccine. The discovery of mutations of the virus, as well as the release of vaccines, pointed toward multiple further potential ways for the context to unfold. A global tapestry was woven through the strategies, not only of the invisible formless enemy but of numerous leaders in different locations acting separately or in coordination, with populations that were more or less predisposed to act in ways that redirected and deflected the virus, on the one hand, or aided its further spread, thereby unwittingly making them the enemy's ally. These potentialities in any one national context were shaped by the guidance and example provided by their leadership, the extent to which populations had experience with previous epidemics, the extent to which societies were unequal, distinguished by wealth and race, and their readiness to see beyond the individual to the impact of one's own action on others and thus to a common entanglement of life within a global community.

Initially the virus succeeded in dividing and defeating human populations. The politics of division preceded its arrival by several years and provided fertile ground for the invasion. As of this writing, the war is not yet over. There are signs of a further reversal of strategy on the part of humans, with a new emphasis, with the election of a new US president in 2020, on uniting to defeat the virus and greater awareness that agency lies not in the freedom to do whatever one wants, regardless of the consequences to others, but rather in consciousness of and deference to others, including nature. Efforts to coordinate a global response not only to the pandemic but also to climate change were ramped up, along with greater attention to inequalities of race and wealth exposed by the pandemic. Addressing these challenges would be crucial to the outcome and the question of whether COVID-19 was to be only a first round in a series of ongoing insurgencies, as the virus further mutated, or was ultimately defeated through a combination of the vaccine and greater human awareness of entanglement with other humans and nature.

The development of a vaccine and a major mobilization to inoculate publics provided light at the end of the tunnel. At the same time, new mutations of the virus appeared, raising questions about the extent to which the vaccine would be able to resist them. Some, such as the biologist Frederik Leung, who studied the SARS coronavirus in Hong Kong, argued that vaccines actually encourage further mutation, whereas the virus might naturally burn out if allowed to take its natural course (Machphail, 2014: 194). Voelkner (2019) argues that if we were to ride the *shi*, that is, to allow the potential of the situation to develop of its own accord, this might mean living harmoniously with the virus.[17] The hope was to inoculate a sufficiently large percentage of populations to create herd immunity. But coordination of what is necessarily a global effort proved to be difficult as 'vaccine wars' developed around the global distribution of the vaccine and wealthy nation-states in particular hoarded supplies. The distribution of the vaccine in zones of military conflict was a further challenge.

The formless enemy continued to mutate as it competed to outwit the development of vaccines. While vaccines are important, they will never be enough if humans fail to see their role in creating the conditions for the virus to prosper, not least in how they live and interact with nature. The virus was a response to environmental change for which the human 'we' is

[17] The author published this article before the pandemic and may or may not have a different view on this since.

responsible (Settele et al, 2020). Returning to 'normal' life, that is, life as it was before the virus appeared, would only enhance the probability that the cycle of pandemics would be ongoing. Defeating the virus would ultimately depend on befriending it and learning to live as part of nature rather than seeking control. From a Daoist perspective, to control is to lose control. Rather than acting on an object, the strategic actant necessarily negotiates relationships and context, navigating a path between self-differentiation and sensitivity to the impact of actions on others. Wearing a face mask became a symbol of this relationality in the West, as it already was in Asia. A further potential would be to reconfigure the war, pitting states against states, as arrogant leaders will always try to do. Changing the 'cut' from humans versus the virus to humans against humans, perhaps even wielding military artefacts, would mean that 'we' all had lost. The new devastation wrought by military conflict on the environment would further destroy the nature that forms the habitat of life.

SECTION III

Entanglement and Karma

SNAPSHOT 5

Navigating a Participatory Universe

> Without performing action man [sic] is not free from the bondage of action.
>
> *Bhagavad Gita*, Karmayoga 3(4)

The fundamental shift that emerges from the snapshots thus far is from 'being' or 'thing-ness' and the pursuit of truth (indebted to Newtonian science) to 'doing', which relates to experience and action in the world. The latter is more characteristic of quantum theory, Buddhism and Daoism. Buddhism and the Daoist 'acosmos' do not rely on divisions into 'levels of analysis'. The latter practice has played an important role in the constitution of international relations as a separate field of study. In its place we see a holographic relationship between parts and wholes, with overlapping and extending circles in continuous motion, where self, mind, body, language, sentience and emotions all matter, not only for humans but for all life. While not constituting a theory of everything, the snapshots do suggest that everything is 'in the soup' (Ling, 2013: 99). The world is in motion. Things are always happening. Change and impermanence are the norm rather than the exception. As Ames and Hall (2003: 13) state, the Daoist correlative cosmology begins with the assumption that 'the endless stream of always novel yet still continuous situations we encounter are real, and hence, that there is ontological parity among the things and events that constitute our lives.' Buddhism and Daoism provide insight into how the actant might, in the midst of swirling motion, act from a place of calm while maximizing the potentials of a situation. The *Daodejing* advises against trying to impose control over the motion as the result will be a loss of control. Nature will, on its own, re-establish a balance if humans would only stop inundating it with toxins.

A particular type of human artefact, which emerged with the industrial revolution and the development of science, sought control over nature to the end of allowing human flourishing and progress. With the emergence of the Anthropocene, this control is almost complete. Control of nature, as Cheng (2018) argues, goes hand in hand with the political domination of human beings. The issue of control, or the desire to impose control, suggests that, within the context of modernity, it was less radical uncertainty that required navigation than a structure that limited movement. The agent/structure problem is closely linked to questions of free will and determinism, which within international relations begin with the 'thing-ness' of states. Wendt's (1992) seminal article 'Agency is what states make of it' asked whether states can exercise agency vis-à-vis a structure of anarchy. Wendt's (2015) later work, from a quantum perspective, acknowledged that constructivism could not account for mutual constitution because it is indebted to the 'thing-ness' of a Newtonian metaphysics. The question, then, is how to reposition the apparatus in such a way that further opens up the black box of the state while also transforming the either/or relationship between agent and structure into either/and such that, as Wendt (2015: 257) notes, entangled parts and wholes can be understood to be 'co-emergent', rather than the whole being seen as arising from ontologically prior parts.

The objective of this snapshot and the next is to explore action within relational entanglements of power that extend across time and space. Both snapshots zoom in on a concept of karma, but from different angles. Karma means action, but is more than a relationship where one thing impacts on another. Action arises from within relational entanglements that are non-local. Karma involves an activation that can reverberate into the future, reproducing extant structures of power and/or changing them. This snapshot explores the enactment of self and the activation of change within a structure of empire. The next snapshot focuses more on the level of systems and explores a notion of karma as habit. The one zooms out from the individual actant, in this case Gandhi, to the context of an independence campaign which sought to disentangle India from the British empire; the other zooms further out to a systemic relationship and the reproduction and transformation of a particular entanglement across time. The two together provide a different understanding of action than that provided by the agent-structure debate.

India, a five-thousand-year-old civilization named after the Indus River, had a long history of trading with Europe, Asia and the Ottomans. A multi-cultural civilization with its own rich traditions of art and science was, however, over time transformed as a vast undivided geographical territory was subjected to various layers of intrusion by Europeans in search of spices and other goods. The first European to arrive was Vasco de Gama, a Portuguese explorer, in 1498. In 1600, the British East India Company was formed to follow the Portuguese into India. In 1602, Dutch merchants followed with the United East India Company of the Netherlands. The British were more cautious and gradually established themselves as peaceful traders, while the Dutch and Portuguese waged a war of attrition. In the early nineteenth century, the British continued commercial development in India and harnessed the country to its world empire.

Britain was weaker, economically, politically and militarily, than India in the seventeenth century, but gradually transformed India's economy to benefit its own global empire (Eacott, 2016). The biggest transition came with Britain's industrial revolution. The invention of the cotton gin in the United States in 1793 and the mechanization of spinning and weaving in the Lancashire textile mills of Britain from 1815–40 priced Indian textiles out of the world market (Marks, 2015: 102–3). India lost many of its overseas markets in textiles as its own price levels were lowered, after which it suffered a long period of depression. Direct revenues from the sale of opium produced in India and sold to China financed the colonial administration while also bringing large quantities of silver back to Britain to underpin the global domination of the pound sterling until World War I (Hervia, 2003: 313). By the mid-nineteenth century, the development of railways and steamships connected India again to a world economy, but in a very different way than had previously been the case. India had always been an agricultural country, for instance, growing cotton for its textile industry, or, under the Mughal Empire, growing opium for domestic use. The British Empire reduced India to an export economy for British profit.

The British colonizers, whether in the Americas, Africa, the subcontinent or the Far East, changed the game in these locations. Local knowledge systems were either discredited or erased, and were written over by English language, law and custom. With this imposition, distinctions between overlapping multi-cultural and coexisting populations were transformed into points of division which could be manipulated in the pursuit and maintenance of power. From a purely materialist worldview, the party with the most military or economic power, in this case Britain, had a definite advantage. We want to shift the focus from a purely material power relationship in order to look at the dynamics of Gandhi's nonviolent strategy in the transformation of India's relationship to the empire.

Resituating the strategist from a world of pure matter, where material power is everything, to a quantum world of impermanence and change impacts on the reality that is seen and engaged. Gandhi and the roots of his strategy in, among other texts, the *Bhagavad Gita* acquire a particular meaning when placed against a quantum backdrop. But the relationship goes both ways. Gandhi's strategy also provides insight into the American theoretical physicist John Wheeler's claim that we live in a 'participatory universe' (Folger, 2002) where observation and action shape the world. To set up the argument I first contrast a conception of 'things' and 'actions' as they relate to measurement and structure. The second section brings this contrast to the concept of desireless action, which has a central place in Gandhi's strategy. Arti Dhand's (2018) critique of *karmayoga* provides a foil for exploring the significance of the claim that we are participants in creating reality rather than determined by it. The third section looks more closely at Gandhi's emphasis on the 'fruits of action' and how this relates to his unpacking of the relationship between self-reform and desireless action as a point of departure for political change. The fourth section brings the categories of unseen and seen to an analysis of the transformative impact of colonialism in India, and examines their place in Gandhi's strategy of nonviolence (*ahimsa*). The final section returns to the notion that we are all participants in creating reality.

The previous snapshot explored strategy as a tool of navigation in the context of war. Sun Tzu's objective was to minimize the need for recourse to force. The Hindu *Mahabharata*'s theme of navigating the universe without doing damage or causing harm (Datta-Ray, 2015: 135) resonates with Sun Tzu's Daoist strategy. Gandhi's experiment went beyond Sun Tzu's efforts to avoid war to explore the potential for nonviolent engagement with an other to the end of transforming relationships. Nonviolence (ahimsa) is consistent with a goal of navigating the universe without doing harm, and rests on a much different strategic dynamic than the imposition of force with violence.

Things and actions

Two points require consideration in moving from the analysis of 'things' as material objects to the analysis of 'entanglements' from which actions arise. The first regards the nature of measurement. If the world is composed purely of material things, the scientist, the apparatus of measurement and the objects to be measured all exist in their separateness. Things have a location in time and space, and measurement consists of quantifying the distance between objects. The objective is to identify patterns and regularities that structure reality. Reality is assumed to exist independently of the observer and thus remains unaffected by its measurement. The central concern is the degree to which the measurement provides a true depiction of the world as it is.

The measurement of quantum phenomena is more complex. The world is not static but in motion. What is measured is thus less what 'is' than what 'might be', given multiple potentials. There is no single angle from which the world or any other object can be measured in its completeness. The measurement takes place from within the motion, which means that the apparatus and the observer move together, shaping the angle of observation, what is measured and what is seen. The apparatus, the observer and the observed phenomenon change alongside one another. The measuring apparatus enacts a 'cut' which is an 'intra-action' from which separation and difference emerge (Barad, 2007: 140).[1] Gandhi, like the scientist, considered himself to be engaged in an experiment (Adam, 2006). Gandhi and the other actants in his experiment are inseparable from the measurement itself. In this respect, action is an enactment and activation of a measurement that is bound up in the constitution of a world.

The second point regards the nature of the reality to be measured. The 'reality' in question is not only in motion but may not even exist until an observations is made! As John Wheeler famously stated: 'No phenomenon is a phenomenon until it is an observed phenomenon.' There is a debate within quantum physics regarding the extent to which particles and waves constitute a reality at all or are merely a function of the physicist's mathematic equations. As Ball (2018: 99) states: 'There is a shared desire to regard Schrodinger's wave function as a physical object itself after forgetting, or refusing to accept, that it is merely a mathematical tool that we use for the description of the physical object.' Quantum theory reveals the illusion at the heart of classical physics, that is, the assumption that reality is composed of fixed particle states and that we can somehow get beyond the tool of description to the object itself (Ball, 2018: 100). The experience of the observer is instead bound up in the production of knowledge of phenomena.

The third point regards the significance of motion and the illusion of substance for how structure is understood. The problem of structure begins with a tendency, even among quantum physicists, to cling to an illusion of substance. The illusion has found expression in a common error. The error rests on an assumption that reality represents a stable state that may be 'disturbed' through a clumsy process of measurement (Ball, 2018: 68). Change, in this error, represents a deviation from a background composed of pre-existing properties, and arises from a blunder by the measurer. Quantum measurement reverses the relationship expressed by the error; properties emerge from the measurement itself as potentials collapse into form.

In the classical view, change is an aberration from the stable properties of an independently existing reality, which, as structure, determines and limits what is possible. By contrast, quantum measurement involves a wave function collapse from which physical properties emerge from indeterminism and change. The former emphasizes matter and form, and structures are understood to be real. The latter is concerned with potentials and the emergence of form from formlessness. The appearance of structure rests on an illusion of permanence, and is thus both real and non-real. The classical model seeks truth in an independently existing reality. The quantum apparatus

[1] Interaction assumes an exchange between separate parts in which they remain unchanged. Intra-action, by contrast, begins with the whole and the constitution of separability as boundaries are drawn in an active process.

measures a process of becoming and the generation of knowledge that is holographic. The measurement problem relates not to the disturbance of an a priori system arising from a blunder; rather the system has no particular properties or character *until* a measurement is made. Different measurements will produce different realities that aren't necessarily compatible.

Measuring human intra-actions

The juxtaposition of scientist and strategist is not to suggest that quantum science provides an analogy or metaphor for thinking about strategy in the world. Bohr's (2010) parallel suggests that science and strategy may be two different manifestations of action that produces the world. Political strategy is a domain which, as Bohr suggests, might be elucidated through the parallel. Both the scientist and the strategist are bound up in the phenomena to be measured. The one draws on a mathematical language to model microscopic intra-actions. The other relies on everyday language to navigate the macroscopic world. Within a quantum state, systems have no properties until they are measured, and different measurements will produce different realities. As Wendt (2015: 217) states, 'In language what brings about a concept's collapse from potential meanings into an actual one is a speech act, which may be seen as a measurement that puts it into a context, with both other words and particular listeners'. His argument about language extends to social structures, which, he claims, 'are, physically, superpositions of shared mental states – social wave functions' (Wendt, 2015: 258). In contrast to a materialist view of structure, these structures are potentialities, which like the wave function are unobservable until they collapse into a particular form. These structures are people and their practices, and thus express consciousness as well as materiality As shared mental states they are ontologically emergent from the entanglements of the agents who constitute them.

Measurement is thus bound up in what can and cannot be said about a particular phenomenon, in this case India. The British measurement of India as the 'jewel in the crown' of Queen Victoria, that is, as her possession, contrasts with Gandhi's measurement that brings India's long cultural and economic history into view.[2] The apparatus of measurement is language. The observer is entangled in language which impacts on what is seen. Queen Victoria or Gandhi's words might be understood to provide two distinct narratives of India, which they indeed do. However, framing the problem as one of quantum measurement clarifies why these narratives are not simply

[2] In retrospect, the articulation of India as the 'jewel in the crown,' is ironic given that the jewel in question was the Koh-i-Noor diamond which was stolen from India (Boissoneault, 2017).

representations of a world that is assumed to exist independently of it. The process is bound up in the constitution of worlds. Because the system has no particular properties until it is measured, the observer participates in its creation, reproduction or transformation.

From here we return to the contrast between the perception of change as an aberration from an otherwise stable system versus change that begins with an act of measurement, which brings about the collapse of a different potential. The contrast relates to two different notions of social reality. In the classical view, a clumsy measurement or mistake gives rise to a deviation from what would appear to be a stable system or structure. The stability of a social system is due, for instance, to patterns of social stratification which inform roles, norms, laws and so forth. The latter are self-reproducing as they become habitual. Any deviation from the structure will appear to be a disturbance to an otherwise functioning system. Action arises from a position within the social structure and is thus more or less predetermined. Within the caste system of India, or, for that matter, the structure of British rule, social order prescribed roles that would determine or constrain the possibilities of the occupants of any one position within it. Social structure had the appearance of substance and of representing reality 'as it is'. While real in its effects, the structure was a habitual manifestation of a particular hierarchical system.

A Brahman had different social capabilities and roles than an Untouchable. Under imperial rule, British civil servants occupied different roles and positions than Indian subjects. In so far as the latter did hold positions within the empire, they would reproduce the rules and norms of British colonial order. An Oxbridge education, wearing British clothing and so forth, which were available to a small Indian elite, contributed to the production and appearance of stability and acceptability. A political strategy that sought to disrupt this structure would be considered a disturbance or an aberration of an otherwise normal order. The agent disturbs, rather than reproduces, and is thus positioned outside and viewed as a terrorist, rebel or criminal. From the British perspective, Gandhi, at different times, was so categorized, as exemplified by the very long periods he spent in prison for disrupting the order of empire, even while he himself was a product of India's elite.[3]

Within a quantum world, to frame Gandhi as the agent who disturbs a pre-existing order is not incorrect, but it is incomplete. A phenomenon cannot be understood to be a manifestation of itself, that is, a disturbance of 'what is'. At the very least 'what is', as a conventional phenomenon, is temporary

[3] As a member of a higher caste, Gandhi had greater potential than, for instance, a *shudra*, or member of a hereditary lower labouring caste, to occupy the position he did within the freedom struggle, and eventually became 'father of the nation'.

and subject to change. Its permanence and solidity is a mere illusion, like the individual experience of 'I'. From the perspective of Victoria's jewel, Gandhi is a disturbance; however, Gandhi, as a strategist, enacts a different world. His actions are manifestations of particular potentials, which differ from those of the Queen's jewel.

From the perspective of 'what is', Gandhi's space for action and movement is both enabled and constrained by the position he occupies as a member of the Indian elite, or, alternatively, as a disruption to the structure of British empire. The choice may appear to be either/or, but from the perspective of quantum worlds, Gandhi is both. Gandhi is a product of his dependent origination within the structure of, for instance, the caste system, which is part of the world he seeks to transform yet from which he cannot entirely disentangle himself. He acts as if a different reality exists and thereby participates in its manifestation, which necessarily means a transformation of himself. As he ([1913] 1964: 158) stated: 'If we could change ourselves, the tendencies in the world would also change. As a man [sic] changes his own nature, so does the attitude of the world change towards him … We need not wait to see what others do.'[4]

The stable extant order is disturbed by Gandhi's action, but the disturbance is secondary to an action that arises from a repositioning of the apparatus and thus a different vision of the world, and of the self within the world. The enactment is a function of position, and of a particular way of seeing. It cannot be separated from the action or the fruits of that action. In the classical view, intention and consequence are separate, which means that one or the other must lead and take priority in deciding how to act. In Gandhi's view, an intention, such as sacrifice or egoism, is embodied and entangled with the act, which is consistent with his view that means justify the ends. The contrast remains abstract but hopefully will take on more flesh in what follows.

Desireless action

Desireless action, a central theme of the *Bhagavad Gita*, is action that is detached from the fruits of action. Action should be without attachment or desire for its consequences. Desireless action might in the first instance be placed in the framework of Max Weber's ([1919] 2004) famous contrast

[4] The more frequently quoted phrase 'Be the change you want to see' has been the subject of bumper stickers but also some controversy, not least because it suggests that it is possible for individuals to change the world. It has a much different meaning when taken out of an individualist framework and placed in the relational perspective of Gandhi, as will later be explored.

between the saint and the politician. Weber constructs a distinction between the saint, who acts with good intention regardless of the consequences, on the one hand, and on the other, the politician, who must always keep in sight an outcome that expresses the greatest good for the greatest number. Intention and consequences, in this argument, are separate, and one or the other must have priority. Weber tries to clarify why the politician may sometimes have to act with less than ethical means if required to achieve a good end. His framework assumes a linear means-end relationship between action and consequence in this world, where outcomes can be measured against a near future. There is some resonance between Weber's depiction of the saint and Gandhi's desireless action. The problem with the saint, from Weber's perspective, arises from the greater emphasis on intention and the means rather than the ends of action. The road to hell is, according to a popular expression, often paved with good intentions. Achieving good ends may, in certain circumstances, require less than desirable or ethical means.

Weber's concerns are justified within a linear means-end relationship where action and consequence are understood to be separate and separable within a cause-effect relationship. However, to read Gandhi's action through Weber's saint example is to obscure the extent to which he was acting within a different and non-linear world of entanglement. Buddhist or Hindu concepts of karma (which means 'action') allow for the possibility that the fruits of action will only be evident at a much later time, whether in this life or another; indeed, the fruits may shape the very question of whether another life becomes manifest. The concept is, in both traditions, inseparable from notions of rebirth and a more cyclical concept of time, where past, present and future are entangled. As the concept of karma ties the snapshots of this section together, I separate out two aspects of the concept. The first, to be discussed here, regards the link between desireless action and karmayoga, which expresses a parallel to the quantum notion that we are participants in constructing the universe. The second regards the relationship between karma and time, which will be a focus of the next snapshot.

Karma has meaning within a cluster of concepts, including for instance, *dharma*, rebirth, *himsa* and *ahimsa*, although the constellation of meaning changes somewhat as one moves from Hinduism to Buddhism to Jainism or Daoism. A detailed analysis of these differences is not the objective here. The objective is instead to explore the parallel between the concept of karma and Wheeler's claim that we are participants in constructing reality and what this parallel suggests about the nature of action in the world. The concept of karma in popular usage tends to place less emphasis on its meaning as action and more on the notion that one's present experience is a consequence of previous good or bad behaviour. Any one action may circle back on the agent. The fruits of action are built into the concept from the start, although, distinct from the popular usage, karma refers to the sum of actions across lifetimes, and not purely to individual action. Consistent with the larger theme of emptiness and form, a key question then becomes whether the action and consequences are manifestations of conventional order or a more universal truth and morality, or some relationship between them. The question points to distinct meanings of the concept of dharma. Dharma can, among other things, refer to one's individual duty or to a notion of moral law. The key issue in the context of Gandhi's action is the extent to which the two are in alignment, and in particular, the extent to which conventional order embodies the universal.

Karmayoga and desireless action

Karmayoga, the 'yoga of action' (Raju, 1954: 210), is one of three spiritual paths within Hinduism. It is the path of unselfish action, which involves both purification of mind and action in the world.[5] According to the doctrine of karma, present circumstances are determined by past deeds and present actions similarly determine future circumstances, which suggests some kind of moral causation across and through time, past, present and future. However, as Chatterjee (2018: 55) notes, if present actions are potentially determined by action in a previous life, one can hardly be said to be morally responsible for one's actions. Nor can one be considered an agent or a participant in constructing reality. Having said that, determinism in this context need not refer to the extremes of fate or fatalism, concepts which are rejected by most philosophical systems as incoherent and inconsistent.

Fate and fatalism suggest a relationship to a transcendent force that controls movement in the now. The more anodyne, albeit tangible, form of determinism is conventional in nature; that is, societies that are based on rigid hierarchies that limit movement, particularly of those at the bottom. Action that is determined by fate is of a different kind than action that is determined by, for instance, a master-slave relationship wherein the former has the conventional power to punish the latter for actions that deviate from the social order as established by the former. The distinction is complex. If observations of the universe are always limited by the positioning of the apparatus of observation, then the frequent conflation of a highly regarded worldview, such as that expressed in the *Mahabharata*, with ultimate truth is to be expected but may be problematic in practice. Particular interpretations

[5] The focus on merit or demerit, reward or punishment, distinguishes the Hindu understanding of karma from the Buddhist. Reward or punishment suggest a superior power that dispenses justice, which is more easily accommodated by a God-centric religion like Hinduism. Buddhism instead associates karma with laws of nature, which will be explored in more depth in the next snapshot. Both resonate with a quantum understanding of entanglement.

expressed in conventional form may become confused with or treated as equivalent to universal truth, of which they are only approximations. The *Rig Veda*'s claim that truth is one but has many names captures the distinction.

Dhand (2018) associates desireless action and karmayoga with a determinism that reinforces the status quo, which he maps on to Gandhi. He argues that karmayoga is a remnant of a Brahmanical worldview that prioritizes order over ethics. Further, the ethics of the *Mahabharata* are 'robbed of their teeth by the conservatism of the *Mahabharata*'s vedic commitments, which are structured upon and hold tightly to a hierarchical model of society' (Dhand, 2018: 87). As a consequence, karmayoga, along with its correlate *svadharma* (one's duty within a social order), both advocates and entails an abdication of moral consciousness (Dhand, 2018: 94–5). Dhand's point is that the Brahmanical worldview uses the doctrine of karma to justify the caste system, and in so doing, loses its ethical compass, which presumes a loss of ethical agency. Karma is conveniently used to argue that those who are high within the Indian caste system have been rewarded for action in another life, while those at the bottom, not least Untouchables, are punished through no fault but their own. The rich and powerful deserve their wealth and position, just as the poor or weak deserve their lack of both.

Dhand (2018: 85) argues that karmayoga is a 'technique for cultivating soteriological liberating habits' that promote self-discipline, equanimity and calm contentment with one's lot; that is, passivity. Paradoxically, the practice cultivates liberating habits that are useful in the maintenance of social order. Karmayoga benefits the self through acts of mental discipline, while making one better able to function and honour one's responsibility within a social ecosystem (Dhand, 2018: 84). The central figure of the *Bhagavad Gita*, Arjuna, combines his duty as a warrior with calmness, equanimity and non-attachment, and thereby 'achieves the renunciation and serenity of a sage without making oneself useless to society' (Dhand, 2018: 84). Karmayoga is ingenious, Dhand (2018: 83) argues, allowing Brahmans to package world renunciation in the form of world engagement – a fail-safe 'middle path' between renouncing the world and being its victim. Dhand sums up the significance of karmayoga in the *Gita* as follows: Don't do what you want but what you ought; disengage the motive for action from its expected reward and act in a spirit of non-attachment, motivated only by duty.

Gandhi, who drew on the *Bhagavad Gita* to reflect on his strategy, emphasized desireless action and motivation by duty. He did also promote 'liberating habits'. Dhand extends his critique to Gandhi, but it is somewhat more difficult to grasp how the latter's leadership of the independence movement could be seen to maintain social stability, and even more so how his actions could be understood as determined, not least because many of them were deemed to be illegal within the prevailing social order. His campaign undermined the social stability of the British Empire, but traditional Indian structures of authority were by no means let off the hook. His challenge to the exclusion of Untouchables represented an attack on more traditional Indian structures. At the same time, his attempts to ameliorate conflict between Hindus and Muslims, in the context of the modern nation state, harked back to an earlier experience of empire in which multiple ethnic groups lived side by side. Indeed, the category of 'Hindu' was a British invention (Mishra, 2002). Gandhi's use of civil disobedience suggests that the duty of the individual should be consistent with universal order in the first instance, which represents a somewhat different take than Dhand's argument. If civil duty is inconsistent with a more universal moral order within which ahimsa (nonviolence) and truth are intertwined (Gandhi, 1950: 251), then action that is contrary to the laws of the conventional order would be justified.

Dhand's critique might be reinforced by the frequent assumption that nonviolence (ahimsa) is passive in contrast to violence (himsa), which is more often assumed to be active, combined with a further assumption that the latter is more effective because it presents a greater threat to social stability. However, claims about the nature of nonviolence often reveal a lack of understanding of the conception of power upon which it rests. Gandhi saw karmayoga as a remedy for Hinduism's over-emphasis on spirituality precisely because the former encouraged a life of engaged and energetic activism. Karmayoga represents a middle way between transcendence of and presence in the world, which is a tension that is expressed in the *Mahabharata*. Gandhi's concept of desireless action was neither passive nor lacking in strategy. Dhand (2018: 88), by contrast, suggests that the contradiction between transcendence and action in the world is resolved once dharma is identified with duty, and one's individual duty with svadharma. So when Khrishna instructs Arjuna to do his duty as a warrior, he refers to his class. In this argument, individual duty is part of an encoded conventional system rather than an exercise in moral judgement. Arjuna merely identifies the aspect of his identity that is relevant for the occasion and mobilizes it (Dhand, 2018: 89). Fulfilling individual duty as a member of a class, whether Brahman or Untouchable, is fulfilling dharma.

The tension between the transcendent and the imminent rests on further layers of complexity as well as a specific notion of agency. If things are always happening and we are always acting, the question is less one of our being determined than of the extent to which action arises from the multiple potentials that are available at any one moment in time. The further question is whether we merely enact social roles or are capable of repositioning the self in order to see that human conventions arise from within an entangled universe but do not in themselves express universal truth. Human artefacts, and the distinctions and divisions they construct, do not reflect the world as it is but are a product of our making. We are not singular but multiple in the potentials that have shaped who 'we' are, whether individual or society, and this multiplicity itself opens up spaces for action.

Gandhi would have felt the pull of his socialization, as would any human, but the force of his action was to enact a relationality of a very different kind than either the Brahmanic or British order, which in the context of British colonialism were entangled. The intention of this action was to reveal the suffering that was constituted by these orders, to make the hidden visible. Conventional duty, whether Brahmanic or British, must, from this perspective, be sacrificed to duty to a more universal moral law, resting on the truth of nonviolence and its ability to expose the structural violence upon which conventional order relied. Indeed, acting with intention is crucial to Gandhi's strategy of 'self-reform', which itself involves seeing 'reality' from a different angle.

Desireless action relies on two different conceptions of self in the world. The problematic nature of self is key to how karma, karmayoga and desireless action are understood. As revealed in the previous snapshot, action arises from both the process of individuation and the whole within which the actant is entangled. Dhand's (2018: 86) more determinist interpretation loses some of the 'powerfully conflicted' nature of the *Mahabharata*. In his argument, fulfilling one's dharma is ultimately fulfilling a pre-ordained role in the hierarchical social forms of traditional Indian society. He suggests Gandhi does the same. In the following, I argue that transforming the self/other relationship was a central objective of Gandhi's strategy, an objective which was not entirely fulfilled. Self-reform involves a transformation that begins with seeing the self from a different angle, which by necessity changes the position and angle from which the other is viewed.

Karma and fruits of action

The concept of karma combines action with the consequences of action, which may rebound in the future. Are the fruits of action best thought about as consequences, and if so, what does this mean? Building on Weber, the concept might refer to the relationship between action and outcome. In this construction, Gandhi's action was not entirely desireless in so far as it began with a strategic objective, that is, Indian independence. But the intention (to free India) and the objective (freeing India) were intertwined, rather than separate as in Weber's argument. Acting freely as an Indian was inseparable from the objective of freeing India.

A further interpretation of the fruits of action ties them more closely to the concept of karma. The meaning of karma varies across traditions. The Buddha, for instance, held a common Indian perspective regarding the existence of natural law (dharma), a physics which was at work in the unfolding of worldly events (Adam, 2006: 6). Karma is an activation that reverberates through time. The opening to the *Dhammapada* captures the idea of a regularity that is law-like and moral, while also highlighting the importance of mind and intention: 'Our life is shaped by our mind; we become what we think. Suffering follows an evil thought as the wheels of a cart follow the oxen that draw it' (Easwaren, 2007b: 105).

In Buddhism, impure states/action are considered unwholesome or unskilful (*akusala*). Unskilful actions are rooted in the 'three poisons' (greed, hatred and ignorance), which stand in the way of our seeing reality as it is. Ignorance is considered to be the fundamental human problem. As such, liberating awareness or wisdom arises when delusion is removed. Illusion refers not merely to an absence of knowledge but rather to the presence of views that provide an illusory awareness of how things are. The most fundamental of these is the illusion of an independent and permanent self. The emotions of greed and hatred arise from this false idea of self. Removing the illusion means also removing self-centred craving and antagonism towards others (Adam, 2006: 7). Actions arising from the three poisons are unskilful because they lead to future suffering, for the agent and for others.

Actions are not impure because they give rise to negative outcomes, as in the logic of consequences, but rather the reverse (Harvey, 2000: 49). The karmic fruit or result is a function of the intention, and in the case of an unwholesome intention, one marked by the presence of the three poisons. As in the passage from the *Dhammapada*, the mental activity of intention is inseparable from mind. But mind relates to something larger than the individual, that is, to the objective moral law (dharma) that is at work in the universe. The effort to align one's personal moral conduct (*sila*) with the working of the universe is considered essential for following the path to awakening, which involves training to be nonviolent in thought, word and deed (Adam, 2006: 7). Someone who has purified the mind and attained liberation embodies moral law effortlessly (Harvey, 2000: 44). An intention that expresses hatred relies on the delusion of self. Higher spiritual realization, by contrast, rests on nonviolence as a natural and spontaneous expression. This highest 'truth' is beyond words, whereas nonviolent action is a part of life and living. The two, that is, truth and nonviolence, come together in Gandhi's strategy of *satyagraha* (truthforce).

Aspects of Buddhism are evident in Gandhi's politics of self-reform. Given that both Hinduism and Buddhism are products of Indian thought, one would expect some overlap between them, and that both would have influenced Gandhi. The biggest tension here regards the place of God in Hinduism, and God's absence in Buddhism. From the perspective of the larger argument of this book, each might be said to use a different language to give form to a truth that is beyond words and ineffable. In a discussion with a group of Theravadin monks, Gandhi (1950: 272) pointed to the 'unalterable existence of the moral government of the universe', and the Buddha's claim that 'the Law was God himself'. According to Adam (2006: 8), Gandhi implies that if an objective moral principle of dharma is at work in the universe, it doesn't matter whether we call it God. The words that are used matter less than the actions that flow from the realization of dharma. Indeed, Gandhi himself was more inclined to refer to the highest principle as 'Truth' (*satya*) rather than 'God' (Gandhi, 1950: 247). Truth is a universal value. Nonviolence (ahimsa) is the natural human expression of truth, and is an effort to embody this state. Ahimsa and truth are intertwined. It is impossible to disentangle and separate them (Gandhi, 1950: 251). Gandhi (1950: 232) accepted that it would be impossible to go through life without causing harm to another and that there are occasions where the destruction of life is necessary. His key point was that action should not be motivated by intention that is marked by selfish interest or hatred.

Within this argument, individual duty (dharma), as discussed by Dhand, is most effortless when it is consistent with universal moral law (dharma). While societies often attempt to align conventionally defined duties with some notion of universal law, there is always a tension, given that the latter rests on a truth that cannot be captured in words. Dharmic duty cannot, by definition, mean that action is determined by conventional roles. If the latter reinforce forms of structural injustice that are at odds with the principle of nonviolence and truth, then the moral order of the universe would have priority. Gandhi viewed nonviolence as the process of embodying and realizing truth, which is the highest duty (dharma) and which is distinctive of our humanity (Adam, 2006: 11–2). Genuine ahimsa indicates not only the absence of harmful intent but also the actual presence of compassion.

Self-reform and desireless action

As in Buddhism and Daoism, Gandhi's thought begins with the tension between two notions of self, which then has consequences for action in the world. The 'true self' can be distinguished from the more illusory self that is associated with the ego. As Ganeri (2012: 25) argues, the true desires – and with them the true self – are said to be hidden by other, false desires. Here again we see a relationship between what is seen, that is, the false self, and what is hidden, the true self. The false self of the ego is impermanent and bound to its own desires within the conventional world. The true self is bound up with an entangled universe, which in Hinduism relies on a more explicit reference to God (that is, Ātman) than is found in either Buddhism or Daoism. While Buddhism revolves around self/non-self, Hinduism relies on a contrast between 'self' with a small s, referring to ego, separate and localized in the corporeal body, and 'Self' with a capital s, which suggests an entanglement between the individual and God or the larger ineffable reality.

Gandhi's reforms began with the egoistic self, including his own. The battle against the British may have been external, but it was defined by the internal battle against the ego. Gandhi (2009: 58) recounts an exchange with Bal Gangadhar Tilak, a leader of the Indian independence movement and an authority on the *Bhagavad Gita*, who cited the verse 'In whatever way men resort to me, even so do I render to them' as proof that the *Gita* endorsed a principle that we should act towards others as they act towards us. Gandhi argued that the verse did not support this claim and that we should not act towards others as they act towards us as we reap what we sow. 'One cannot do evil to others and expect good for oneself', he argued. If a person gives in to attachment and anger, they will reap the fruit of anger. By contrast, one should 'see oneself in the whole world and the whole world in oneself, and act towards others accordingly' (2009: 115).

Action that is motivated by 'True Self' or the entangled Self is action of a holographic nature. If one is part of the whole world and the world is in oneself then harm to another is harm to the self as well. Even the smallest action has reverberations. You hit me and I hit you back means that not only are the two of us damaged but also the delicate and intricate fabric in which the universe is spun as well. Much like a virus, anger and violence spread. Indeed, war is the manifestation of anger toward an other that expresses a collective form. The central point regards the difference between action that arises from 'self' or 'Self' and the place of desire and the fruits of action in this equation. While the fruit of action is an outcome of sorts, it is not a conclusion, as we might tend to think. Fruits of action suggest that something is passed on, but what is passed on is larger than the individual. A conclusion instead suggests a stable endpoint. Fruits of action, as something passed on, whether to those in one's spatial environment or across time, impact on a way of being vis-à-vis others. The crucial question is whether the other is viewed through the lens of one's own ego-focused self-desire or in terms of a relationship by which both constitute an entangled 'Self'. It is a contrast between an emphasis on property, that is, what is mine, and relationality, that is, the vulnerability and potential suffering that we both share or inflict on one another.

A focus on what is mine creates a demarcation from others. The delicate web is woven in the relationship between self and other, which is the continuous subject of negotiation and navigation. Western society leans too much toward the individual, while Eastern society, it is said, leans too much toward the societal. A balance between individual and society is both necessary and difficult to sustain. Action that is motivated by desire arises from the egoistic self. It seeks to maximize the utility of the individual ego, to use the language of social science. While not by definition in conflict with others, and capable in mutually agreeable circumstances of coordination, what is mine is the point of departure for defining the desires and interests of the self, which means that the other or others, defined in opposition to self, are secondary. The 'I' is also the starting point of the three poisons of Buddhism, that is, greed, hate and ignorance. If motivated first and foremost by the egoistic self and the desire for success, career and profit, what is secondary may of necessity be harmed to satisfy the primary beneficiary of action. If motivated by the entangled Self, what counts is the agent's subjective non-attachment. To lack attachment to the fruits of one's action means there is no self-interest or ego intention in undertaking them (Adam, 2006: 10).

Self-reform, as part of Gandhi's campaign, was the point of departure for political change. The egoistic self needs to identify with the relational and entangled Self if there is to be change at the level of society. Action from this position is desireless in so far as it is less concerned with the fruits of action that will come to the egoistic self, that is, the 'I', and more concerned about the maintenance and/or repair of the relational Self. The latter has been harmed by the reproduction of a particular kind of difference, in which the act of separation is violent (himsa) in its preferencing of the needs or property of some over others. As such, Gandhi's campaign sought not only to get Britain to quit India. He also sought a transformation of individuals and of the relational map of Indian society, which was characterized by divisions that profited some at the expense of others. For instance, for the Indian elite, including some within the independence campaign, trying to be British brought a degree of success. However, wearing Western clothing brought harm to those whose suffering stemmed from the destruction of the Indian economy, which was based on the production of textiles, and which was reorganized for British profit rather than Indian. Further, the formal caste system of India pushed groups, and Untouchables in particular, into severely constrained roles that harmed their dignity, while their labour benefited those above them in society. Finally, the division between Hindus and Muslims was a source of conflict and tension that was manifest in violence against the other, to the detriment of a self-governing India. Gandhi's strategy sought to transform the relational map that kept British power intact. Nonviolence was an essential ingredient of the strategy.

Seen and unseen

While the distinction between self and Self has its roots in Hinduism going back millennia, the focus on self-reform was entangled with the structure of British colonialism in India. Ashis Nandy (2009: 24) refers to British colonialism as a 'psychological invasion from the West'. The invasion began with the internalization of Western values by many Indians, which involved a process of reforming Indian personality. The British could not grasp that Indian civilization went back thousands of years, with its own traditions of art and science. The colonial process imposed a clear distinction between

India's past and present, relegating civilized India to the unseen, bygone past. It further imposed a differentiation between colonizer and colonized, which recast the colonized in a negative and dependent version of its previously creative self, while attributing the degradation not to colonial rule but to an Indian culture which carried the seeds of its own downfall (Nandy, 2009: 17).

Colonization instilled habits of seeing informed by what is more often referred to as the 'colonial gaze'. As Aime Cesaire (1972: 57–8) notes, 'The colonizer, who in order to ease his conscience, gets into the habit of seeing the other man as *an animal*, accustoms himself to treating him like an animal, and tends objectively to transform *himself* into an animal.' Gandhi's emphasis on self-reform expresses not only a concern that the colonizer has a particular habit of seeing Indians, whether as animals or children, as incapable of self-governance; the psychological impact on the colonized was to change the way they saw themselves. The latter involved a degradation of the Indian self, to make it more like the colonizer. Being more like the colonizer was in part a function of buying British, which was a form of collective self-harm. The Indian economy was destroyed by colonialism; colonialism involved cultivating a particular desire for Western goods, which made Indians participants in reproducing colonial structures.

In Nandy's (2009: 34) argument, the model was masculine, based on exaggerated individualism and egoism, as well as what E.M. Forster referred to as the 'undeveloped heart' of the British. The undeveloped heart of the British involved a separation from themselves as well as Indians, and an isolation of cognition from affect, which is not unrelated to a potential for violence. Nandy (2009: 49) suggests that Gandhi was engaged in a universal battle to rediscover the softer side of human nature and the more non-masculine side of men who were themselves part of a different habit of seeing. The Western self-concept necessitated control of the 'objective' not-self, which related not only to an external other but to the not-self within the self (id, brain processes, social or biological history) in order to better control and understand the self of ego, praxis and consciousness (Nandy, 2009: 62). In short, if colonialism was a psychological strategy of invasion to claim the self of the colonized based on a different cultural and cosmological concept, then resistance involved a reappropriation of 'I' in the reclaiming of one's own self.

Gandhi's campaign begins with self-reform as the point of departure for political change. Self-reform is necessary for beginning to make conscious choices that will transcend and transform divisions within society. The central objective of Gandhi's campaign was *swarāj*, which means self-rule. This sacred Vedic word, which is far more than political independence in the Western sense, means both self-rule and self-restraint. As observed in the previous snapshot, differentiation is a part of nature, so the point is not that the Western self differentiates while the Indian self does not. Rather the concept of self as ego relies on a more complete separation, both internal and external, whereby what is mine, and the desires attached to it, is foundational rather than arising from an entanglement with life. The latter is only secondary. The separation constructs a hierarchy between mine and yours whereby the former is imposed on the latter rather than negotiated. Seeing the illusion of self involves a reorientation which allows for the tension between 'self' and 'Self', imminent and transcendent, to be sustained.

In this respect, Gandhi's strategy rests on a fundamental critique of modernity, including the reification of violence, widespread among realists, as a 'natural' and thus 'inevitable' feature of human nature (Steger, 2006: 332), an atomic notion of the self as independent and prior to any social relations, a notion of matter as brute and inert, and the desire to control nature, which disengages humans from life within a habitat. As Bilgrami (2008: 63–4) argues, these Newtonian orthodoxies remove any normative constraints on action and abdicate any notion of a first-person – as distinct from a third-person – view of agency. Gandhi rejected the materialism of modern industrial civilization and refused to see the progress of the modern West as an aspirational model (Chacko, 2016: 195). As Priya Chacko (2016: 197) notes, Gandhi sought to disrupt the connection between political power, violence and agency by eschewing the extension of Galilean and Newtonian laws of the physical world to human relations as expressed in assumptions about the 'state of nature' and 'social contract' from which modern political theory arose. Gandhi (1999: 134) embraced the ancient rishis (seers), who had discovered the law of nonviolence in the midst of violence, as greater geniuses than Newton. In contrast to modern civilization, Gandhi sought a civilization of self-discipline that began with the unity and oneness of life as the foundation of truth, nonviolence and tolerance, which would then underpin concepts of political self-government for the many. Individual self-government meant that India's freedom would begin with each individual taking personal responsibility for changing attitudes of intolerance and exclusivity. Individual reform was the basis for social reform, including of Hindu-Muslim religious conflict, the evils of the caste system and economic inequality.

The dialectical relationship between seen and unseen is fundamental to his strategy. After returning from his time in South Africa, and before launching his campaign, Gandhi travelled to the four corners of India to speak with those suffering under British rule. Actions such as the boycott of foreign goods and the Salt March brought the violence and hidden suffering to the surface, where it could not be ignored, and thereby raised questions about the moral legitimacy of British rule. That the marchers were not themselves violent was fundamental to the power of Gandhi's strategy. Once the violence of British rule was juxtaposed with the nonviolent actant in the full light of day, attention shifted from the actant as an aberration and disruption of a deterministic structure to the illusory nature of the structure and of the moral legitimacy of British rule in India. If life is always viewed from a position in social space, the extent to which the strategist impacts on what is seen, bringing the unseen suffering into the open, not only changes the parameters of moral judgement but also the measurements through which potentials collapse into form. The seen/unseen dialectic is central to the strategy, although less explicit than in Sun Tzu's thought. The concept of desireless action highlights the relationship between the egoistic self and the entangled Self, Ātman. The egoistic self, because confined to the corporeal body, is impermanent, and, as in Snapshot 1, often mindlessly concerned only with its own desires, which can then provide a justification for harming others. From a grounding in classical physics, the focus would be on material power and local presence. A quantum repositioning brings the matter-consciousness relationship into view, and with it the dialectical relationship between seen and unseen.

Satyagraha, or 'truthforce', was the core of the strategy. As William Shirer (1979: 84), who reported on Gandhi's revolution in the 1930s, said in his memoirs, '*Satyagraha,* his supreme achievement, taught us all that there was a greater power in life than force which seemed to have ruled the planet since men first sprouted on it. That power lay in the spirit, in truth and love, in nonviolent action.' Gandhi described satyagraha in terms of a metaphor that likened it to a banyan tree with innumerable branches (Iyer, 1973: 265). The trunk consisted not only of nonviolence (ahimsa) but also of truth (satya). The branches consisted of the many forms of nonviolent action, from civil disobedience and non-cooperation to fasting and social reform. In regards to truth, Gandhi warns of the need to continually remind ourselves of our fallibility and limitations. Human understanding is always imperfect and thus incapable of possessing absolute truth (Steger, 2006: 345).

We have to believe in truth but we cannot possess complete knowledge, and any claim to infallibility would be dangerous. Truth is vindicated not by inflicting suffering on others but through self-suffering. Precisely because none of us can see reality 'from everywhere at once', we are not in a position to know the full truth and thus to punish others. Gandhi's philosophy rests on a deeply relational understanding of our entanglement with others. Jonardon Ganeri (2012: 31) refers to the secret teaching (*upanisad*) that 'the self that gazes out from within my body is the same self that gazes out from within yours'. Gandhi (2009: 97), like Sun Tzu, relies on a water metaphor, claiming that 'as ice becomes what it is from water, so we have all come from the same water and shall turn again into that water.'

The ideal of nonviolence has its origins in the realization that when life is full of suffering we should cause suffering to none (Gandhi, 2009: 115). This provided a point of departure for the conviction that it is preferable to transform the enemy rather than to destroy them. The transformation was not to be realized through passivity or by doing nothing but through action that would strike at the enemy's moral sense of gravity by destabilizing the conventional truth and turning the relationship between seen and unseen on its head. The presence of material power shrouded in the legitimacy of moral goodness sustained the illusion of a structural reality. Looking below the surface, the hidden threads of suffering woven through the relational fabric became visible. The power and the suffering were mutually implicated and entangled.

Surprisingly, Gandhi held that given a choice between passive acquiescence born of cowardice and the use of violence, the latter would be preferable (Ostergaard, 1974). In making this claim he was not advocating violence but rather highlighting the importance of action and a conviction about how one's objectives would most effectively be realized. Indeed, as Ganeri (2012: 5) notes, the self *is* an activity.[6] The satyagrahi general sought not retribution but to transform a conflict situation so that warring parties could find a mutual interest in resolution. Satyagraha rests on a combination of withdrawing one's cooperation from a state that has become corrupt – which assumes its continuation is dependent on the support of oppressed populations – and civil disobedience, in which unjust laws are disobeyed in order to register protest.

Self/self

As in Daoism, the life force, or *prana* in Indian thought, has a central place. While violence is the law of the brute, who knows only physical might, Gandhi (1951: 133–4) refers to nonviolence as the law of our species. He further claims that dignity requires obedience to this higher law and strength of the spirit. Power comes from complete control over the senses and the preservation and sublimation of 'the vitality that is responsible for the creation of life. If this vitality is mobilized rather than dissipated by evil, it can be transformed into a powerful creative energy. This power involves control over not only action but thought' (Gandhi, 1951: 97). The satyagrahi general examines himself continuously, listening to the dictates of the inner self and obeying his inner voice. *Brahmacharya*, which relates to the conservation of vital energy, is the source of the inner strength 'to stand unarmed against the whole world' (Gandhi, 1951: 95). Meditation makes it possible to see and acknowledge the three poisons at work in the general's own mind (Nyanatiloka, 1972: 166). With satyagraha, no less than with military warfare, the discipline of the soldier, and their obeying the general's command, is crucial, even when hostilities have been suspended. But in this war there is

[6] Italics added.

no space for violence, in any shape or form, including of thought, speech or deed. There is only space for soul-force, which is universal and makes no distinction between kin and strangers, young and old, male and female or friend and foe (Gandhi, 1951: 78). While some are believed to follow the law of the brute, the satyagrahi begins with the assumption that no one is so fallen that they cannot be converted by love (Gandhi, 1951: 25).

Remaking the individual self, finding one's higher Self within, is inextricable from transforming the world, and is thus fundamental to Gandhi's campaign. The struggle between good and evil, both internal and external, rests on a recognition that evil can only flourish in the world if allied with something good (Gandhi, 2009: 4). Based on Gandhi's principle of noncooperation, the evil system, represented by the British colonial regime, would only be able to endure so long as it received the support of good people. Once this support was withdrawn, the government could not survive (Gandhi, 2009: 4). Likewise, the selfless action that underpins noncooperation requires shedding attachment to the body, or to forms of cooperation that bolster the system.

Gandhi's nonviolence struck at the moral centre of gravity of British rule in India by challenging the appearance of British goodness and their right to rule India, turning it on its head and revealing the structural violence that underpinned their presence. Here I highlight just two types of action in the context of the Indian campaign. One was the renunciation of foreign clothing (*swadeshi*), replacing it with the spinning of one's own cloth and living more simply according to Indian traditions. In so far as India's economy, which had been based on textiles, was destroyed by colonialism, spinning became a symbol of a new India, with the spinner as a participant in its creation. As regards the boycott, Gandhi (1951: 146) was careful to specify that things should not be boycotted merely because they are British and that the boycott extended to all foreign cloth, because its dumping had reduced millions to pauperism. Further, to focus on British goods was indefensible from the perspective of nonviolence as it would be retaliatory and punitive. The intent was for Indians to act as if they were Indian rather than British.

A second form of action required letting go of an attachment to bodily comfort. During the Salt March, which took place from March to April 1930, rows of nonviolent 'soldiers' were bludgeoned by the forces of Her Majesty's Army as they approached the sea to collect salt, one of the most basic necessities of human life. What was first and foremost renounced was action motivated by the desires of the ego, and not least comfort, which was replaced by voluntary suffering, if need be, and compassion. As Gandhi (2009: 90) states, 'Nonviolence will have become direct experience for us … when our whole life comes to be permeated with the spirit of compassion, when nonviolence manifests itself in its true essence'. This relies on a spirit of sacrifice or *yayna*, which requires letting go of the impermanent ego, that is the self with a small s, and seeing the other and oneself as inseparable, as part of a more permanent Self, as Ātman. Satyagraha assumed that suffering would move the heart of an oppressor or opponent even if reason failed to move the head. The strategy was conceived as a practical experiment in employing nonviolence within the political realm.

Conclusion

Datta-Ray (2015: 135) argues that modernity is coded in violence. He examines the influence of the ancient Hindu *Mahabharata*, not only on Gandhi but on Indian diplomacy following independence, and states that it provides a means of 'navigating a crowded cosmos without damaging it', and of negotiating the universe without causing harm. Gandhi's nonviolence is often interpreted as a weapon of the weak. Gandhi (1951: 51) himself claimed that it is in fact superior to the force of arms. Unlike the example of the warrior Arjuna, preparing for battle, or for that matter Sun Tzu's *Art of War*, Gandhi's relational strategy was one of nonviolence, which began with the self-reform of individuals, including himself. As with most experiments, the strategy wasn't entirely successful. The analysis of how and whether it was successful is not the objective of this snapshot, which is rather to explore Gandhi's engagement in an experiment in which he, like the scientist, was bound up with the apparatus of measurement and thus a participant in the construction of reality rather than a neutral observer. Structure, from this perspective, does not determine; the enactment of alternative potentials reveals the illusory nature of structure, and instantiates movement that reverberates across space and time. The act of observation activates a potential and a wave function collapse, which is a disruption of 'normal', but also contributes to bringing other worlds into being. One might further ask whether Gandhi can claim credit for the British withdrawal from India, or at the very least, whether he had an impact on how the British left, that is, as friends rather than as enemies. These topics must be left for another time.

Gandhi embodied his own dependent origination, but was also a participant in the construction of a different India. Krishna (2015: 140) counters the idealization of Gandhi as an icon of peaceful political change, placing his politics at the intersection of race, caste and the international, arguing that his politics reinforced the very same racial and spatial ordering of the international system that formed the backdrop of his life. Yet Gandhi need not be either an icon of peaceful political change or a part of the order he was trying to change. He was both, either/and. Actants are a product of the relationships from which they arise, yet they may attempt with varying degrees of success to move in new directions as they experiment with different potentialities. In this respect, Gandhi represents the enfolding and unfolding of the hologram, by which he as a part is deeply implicated in the whole.

The tensions within Gandhi's context and strategy are arguably reminiscent of those found in the *Mahabharata*. On the one hand, self-reform was about the potential activation of the entangled Self rather than the egoistic self. On the other hand, the objective was to craft Indian independence within a world of separate egoistic nation states. Self-reform reached toward the relational self, while the parts of the international system, that is, states, are products of an ontology of separation, if not in Gandhi's mind then in the twentieth-century construction that is international relations. International relations is based on a logic that too easily becomes Hobbesian, transferring the problem generated by the coexistence of individual egos to the relationship between collective entities.

James DerDerian (1987) elaborated on the international dynamic in his discussion of modern diplomacy. Modern diplomacy, he argued, rests on the alienation of the separate parts, which is prior to the attempt to cobble together a set of rules that would make communication between them possible. The parts are prior to the whole. The ontology of state diplomatic interaction relates to a conception of grand strategy that emphasizes violence and war. Within Western debates there has been a tendency to equate grand strategy with military strategy, the objective of establishing control over others and the system as a whole, while placing a premium on sheer force and technology in doing so (Yuen, 2014: 14).

Both Sun Tzu and Gandhi developed relational strategies with applicability that extends beyond a focus on states or warfare, and within which achieving objectives without recourse to force is a priority. The emphasis is less on controlling a system than re-establishing balance and thus some notion of harmony. Both strategists would seem to highlight Silove's (2018: 27) emphasis on 'grand plans, grand principle and grand behaviours'. From this starting point, Gandhi's strategy may fit uneasily with the broader literature on military strategy but is, nonetheless, unquestionably grand. The grand strategy of seeking India's independence, against the backdrop of British imperial rule, raises questions about any notion that the state, as a repository of community that seeks security, is ontologically prior. The potential for the state to be an object of contestation comes clearly into view when placed in the context of a fight that ran counter to the grand strategy of Great Britain, the global imperial power of the time. India's independence campaign was precisely one about the means and ends to improve India's security vis-à-vis what was implicitly – and eventually explicitly – defined as an external power. In constructing Indian independence, the internal tension between Hindu and Muslim identity was transformed into a relationship between collective selves, each defined by a desire for independence. Gandhi's relational strategy was constrained by the context of nation states. Gandhi's campaign arose from a relational ontology within a strategy of independence. It thus reflects the fundamental problem at the heart of the *Mahabharata*, that is, the tension between self and Self, and how one navigates the tension without doing damage.

The question is whether action within the world can ever transcend this tension entirely. Is a strategy utopian in seeking detachment from concern for the fruits of action or from the ego self, whether individual or state? The tension is expressed in the conflicting desires to, on the one hand, transform the relationship to Britain through nonviolence, and on the other, to establish India's place as a nation state which exists separately from others, where its independence is prior to a more global relational whole. The desire is a very modern one. As stated earlier, nationalism in India was a byproduct of the colonial experience. Gandhi, the experimenter, was entangled in a larger context which both shaped him as a leader and constrained his movements, though without determining them. His measurements relied on the conventional constructs of the international system he inhabited, that is, the nation state, even while he sought to reposition the apparatus and transform the world.

SNAPSHOT 6

What Goes Around Comes Around

The previous snapshot zoomed in on Gandhi's efforts to navigate the universe without doing damage. The question of what ultimately constitutes damage is not straightforward, and can look quite different depending on one's perspective. When protests erupted in Minneapolis following the murder of George Floyd in May 2020, and initially turned into riots, some observers focused on the damage to property. The point of the protests was to highlight damage of a more profound kind: the murder of a Black man under the knee of a White police officer as part of a legacy of damage to African Americans not only in the present but since 1619, when the first Africans arrived in English North America. Likewise, the protesters who somewhat later toppled the statue of Edward Colston in Bristol, UK, were making a point about the damage done by the transatlantic slave trade to generations of Africans and their descendants, and about the hypocrisy of British society for celebrating those who pursued the trade. Others highlighted the damage done by the protesters to the statue as an historical object. The contested attributions of damage, that is, to property, inanimate objects or human beings who breathe and suffer, contained a measurement of value. That the humans in question had themselves been property, often of those whose lives were celebrated in the statues, only highlights the conundrum.

George Floyd's last words as he lay dying, "I can't breathe", brought his humanity into view, and with it the legacy of a history of American enslavement and the transatlantic slave trade. The moment of seeing Floyd's death as a part of this history contains a number of puzzles. The first regards the ethical justification, in the context of the slave trade, for bringing concepts of property and of human life together in the phenomenon of the chattel slave. The second regards the karmic resonance of the historical practice. 'What goes around comes around' suggests that we reap what we sow, whether in this life or another. Arguments that wealth is a reward for goodness, while poverty and other misfortunes are punishment, rest on a logic of karma, whether articulated by Hindus (Dhand, 2018), the Old Testament Job, Protestants (Weber, [1958] 1976), or in more secular discourse (Sandel, 2020), even when the concept itself is not invoked. This type of argument has often provided a justification for structural inequality and violence. From this perspective, karma would be the wrong concept for making sense of historical and contemporary injustice, or of other cases where suffering has been inflicted on particular populations.

The Buddhist concept of karma, while varying across schools within Buddhism, is very different. The popular understanding suggests that an outcome, that is, good or bad fortune, wealth or poverty, is evidence of the merit of earlier action by an individual. By contrast, within Buddhism, karma is inherent in the nature of things, much like a law of physics or indeed a quantum physics of entanglement. It is not a reward or punishment from God (Harvey, 2007: 16). The impermanence and absence of an essential self or soul in Buddhism points to a different understanding of what carries over from one lifetime to the next. It is not the sum of individual actions, and thus the transfer of a soul – an idea more compatible with Hinduism – but rather the reproduction of a common reservoir of ethical standards of living. Karma demonstrates how ethical standards of living become embodied and reproduced within a relational system that is suffused with a particular consciousness. This consciousness may express morality while also manifesting a system conditioned by the poisons of illusion, greed and hate.

The significant and expanding literature on race and economy highlights the constitutive links between the liberal capitalist economy and racialized constructions of property (for example, Bhandara, 2018; Tillie and Shilliam, 2018). The objective of this snapshot is to explore the politico-ethics of entanglement with tools informed by both Buddhist karma and quantum entanglement, particularly as it relates to the property/life distinction exposed by George Floyd's murder. The previous snapshot zoomed in on a question of how to conceptualize action that potentially transforms a seemingly determinant structure. That snapshot was an investigation of how Gandhi, as an actant embedded within structures of caste and empire, became a participant in bringing about a different potential. The same question is relevant here, but the context is different. George Floyd was an 'equal' American citizen rather than a subject of empire, yet, long after the formal institutions of slavery had ended, his murder revealed a stark difference between the lived experience of Black and White Americans.

A metaphor used by the physicist David Bohm (Horgan, 2018) provides a way to think about the problem.[1] A glass barrier is placed in a tank with fish. The fish keep away from the barrier, and once the glass is taken away,

1 Bohm was making a somewhat different point regarding the incompleteness of knowledge and his rejection of claims by other physicists, such as Stephen Hawking, that it would be possible to develop a theory of everything.

they still won't cross over it. The point is to think about how habits continue long after formal structures have been removed. One might say that Gandhi wandered over the barrier in order to recast life in a larger fishbowl. Floyd's murder revealed a different problem, that is, the persistence of racialized practices that are no longer directly encoded in laws, yet which continued to be reproduced through 'habit'. The habits reinforce a systemic reality where, to return to the metaphor, fish on one side of the invisible barrier are still able to control movement within the entire bowl. The Buddhist concept of karma provides a particular perspective on the relationship between life and property, on the one hand, and on the other hand, an understanding of how habits can become systemic while not necessarily determining action. The parallel between Buddhist karma and quantum entanglement, as the apparatus of investigation, provides a lens through which to view the systemic dimensions of race surrounding Floyd's murder. The analysis opens spaces for exploring a holographic relationality, woven through action that reverberates across time and space, which reproduces a Black-White fault line while giving rise to potentials for transformation.

An apparatus of karmic entanglement makes it possible to analyze a system as a whole that is interwoven through its parts. The investigation begins with the distinction between things and relationships as it relates to the apparatus. The second part moves on to consciousness and property, or the relationship between life force and objects, looking, among other things, at how the enslavement of a human being, that is, the chattel slave, who is both human and property, was justified. The third part elaborates the dynamics of karmic entanglement within the cycle of samsāra as it relates to a system of enslavement. Part four returns to the relationship between seen and unseen, and the silences held within this self-generating system. The conclusion of this snapshot then returns to the intention-action-fruits relationship and the potential to transform the Black-White fault line. At the heart of US race relationality, over time and in the present, is a question about the value of human life vis-à-vis the value of property. The conclusion of the book then zooms out again to the more global dimensions.

The literature of international relations has failed to grant race any explanatory agency, even while it has been integral to the birth of the discipline (Anievas et al, 2015). As Grovogui (2001: 427) argued, the ontology of international relations has buttressed Western claims to moral authority as a provider of rules and models to the rest of the world, including Africa. Events in 2020 drew this moral authority into question. The discussion of the 'colour line' has entered into global politics over the past few decades, and it accelerated and unfolded alongside the Black Lives Matter protests in 2020 (Shilliam, 2020). The contestation following George Floyd's murder revealed the vulnerability and impermanence of the lives of Black Americans in a society that values the accumulation of property and profit, from which earlier generations, as property themselves, were historically excluded. Against the backdrop of the COVID-19 pandemic, George Floyd's murder in Minneapolis was a moment of seeing. Starting with the image of a police officer's knee on the neck of a Black man, awareness unfolded of the entangled history of enslavement in the 'land of the free.'

Things and relationships

Classical Newtonian physics and much Western social science assume materiality or 'thing-ness' in a reality that is understood to exist independently of the meaning attached to it. Truth is a function of the correspondence between object and representation. Quantum mechanics and the ancient Eastern philosophies instead highlight the relationship between the entanglement of consciousness and matter. 'Things' do not exist as essences but are instead fundamentally relational. Relationality is a concept that has many different disciplinary homes and is increasingly a subject of dialogue between scholars of international relations in West and East (see Nordin et al, 2019). Pan Chengxin (2020) identifies several weaknesses in the debate, including the lack of a clear conception of relations and relationality, a tendency to give temporal priority to relations over entities and a failure of the literature to engage with the 'quantum turn'. Relationality, as a quantum phenomenon, he argues, is holographic. At its most basic, quantum holography suggests that an object is both part of the whole and yet contains the whole (Van Daele, 2018: 651). Quantum holography, as articulated by the physicist David Bohm, goes beyond the 'ontological chicken-egg dilemma' of whether relations or things come first, by showing how relations are embodied in things. 'Relations-*in*-things are implicate relations that can be better understood through quantum mechanics, whereas things-in-relations (relations-*between*-things) represent more classical understandings of relations' (Pan, 2020: 3). It is the difference between what Barad (2007) refers to as 'intra-action', by which each party to an engagement is fundamentally shaped by the 'cut' through which their relationship is formed, and 'inter-action', which leaves the separate parts unchanged. The one is mutually constitutive, which as Wendt (2015: 260) claims, is only tenable within a quantum framework; the other is transactional. As Barad (2007: ix) states, 'To be entangled is not simply to be intertwined with another, as in the joining of separate entities, but to lack an independent, self-contained existence. Existence is not an individual affair. Individuals do not pre-exist their interactions.' Parts, including individuals, are always already bound up in the whole, just as the whole is in the parts. Relations are the condition for the being of things or actors.

The distinction between things and relationships connects to a particular understanding of universe and world in which the microscopic and macroscopic, including humans, are interwoven within a universal whole. Bohm ([1980] 2002: 2–3) starts his classic work with the point that individual human beings have been fragmented into a large number of separate and conflicting compartments, not least categories of race, gender, nation and class, including the separation of humans from nature. The fragmentation has its origins in a way of thinking about things as inherently divided, disconnected and 'broken up' into smaller constituent parts, each of which is considered to be independent. A further habit is to take the fragmentation as a description of the world 'as it is'. In a Newtonian world, all is matter and consciousness has no role. Bohm ([1980] 2002: xi, emphasis added) asks the question of how to think about 'a single, unbroken, flowing actuality of existence as a whole, which contains both thought (consciousness) *and* the experience of external reality as we experience it.'

Within the implicate order, everything is enfolded (that is, folded inward) in a quantum interconnectedness comprising the whole universe, which is the fundamental reality, or what Buddhists refer to as 'emptiness'. The implicate order is unbroken, with enfolded structures that interweave and interpenetrate throughout the whole and infuse the broken, fragmented conventional world. What might be presented as an either/or choice, that is, Newtonian or quantum ontology, might then be replaced by the interconnectedness of parts and whole, within which the two orders are mutually implicated. The broken, fragmented classical world represents a parallel to the Buddhist cycle of life and death, samsāra, in which divisions and fragmentation are a manifestation of human illusion. The need to understand the self as separate and containing an essence arises side by side with an unwillingness to see either the dependence of the self on others, that is, one's fundamental relationality, or the existence of suffering in the world.

Actions, in the Buddhist conception, are not discrete movements in local space and time but have implications that enfold and unfold into the future across generations. Karma means not only action but an activation of emotion and memory, linked to a concept of time by which 'what goes around, comes around' (Baldwin, 1985). The cause-and-effect cycle involves the instantiation and assimilation of past intent and its effect and affect in future relationships. Breathing and movement are a part of history and future-making. The action, activation, which cascades across time, is what quantum physicists would refer to as a 'wave function collapse', from which intention and thought become manifest in the world. If thing-ness represents static being, action, activation and wave function collapse are about becoming. The basic issue is how, from a vast field of potentials or wave functions, particular potentials collapse into form. Far from nothingness, as is often assumed, Buddhist 'emptiness', like Bohm's implicate order, is the field of multiple potentials which, in collapsing into form, gives way to the dependent arising of relational selves, shaped by DNA, family, culture and so forth. Consciousness and form are mutually implicated and always exist in a relationship to each other.

The implicate order, like Buddhist emptiness, represents a more fundamental and unbroken reality that *enfolds* all of the potentials contained in memory. The explicate order, like Buddhist dependent origination, *unfolds* particular potentials into the specific detail and divisions of any one time and place. In the unbroken implicate order, the past exists as traces of intra-actions and relationships, of memories and habits which may remain alive and infuse 'relations-in-things' within the explicate order. Barad's (2010: 240) discussion of the relationship between memory and world is relevant here. The past isn't simply there, a thing or an event; nor is the future simply that which unfolds. Rather, as she further states (Barad, 2010: 260), past and future are 'iteratively reworked and enfolded through the iterative practices of spacetimemattering'. Memory then becomes the pattern of 'sedimented enfoldings of interative intra-activity', which is 'written into the fabric of the world' (Barad, 2010: 261). Memory traces are held in the world, which '*is* its memory'. The traces of all the reconfigurations of patterning are 'enfolded materialisations of what was/is/to-come.'

Memories are the 'seeds' of consciousness from which, to use the language of Buddhism, karma arises. Karma is often compared to a seed. Two words for karmic result, ripening (*vipāka*) and fruit (*phala*), flow from this metaphor. According to Harvey (2007: 17), the intention (*cetanā*) determines the nature of the karmic seed, which is the impulse behind an action which activates a chain of causes that culminate in karmic fruit. Good and bad karma are constantly generated in the mind as it attends to and responds to objects of the senses, memory or imagination (Payutto, 1993: 6–8). Karma can take the form of bodily, vocal or mental acts. Karma is enfolded in action and unfolds from it.

Just as the explicate and implicate orders are mutually intertwined, so are Buddhist emptiness and dependent origination. Both give rise to relations-in-things, even while the illusion of separateness is maintained. The second-century Buddhist philosopher Nāgārjuna wrote millennia before the emergence of the modern world in the seventeenth century. In the time in between, and across global space, there have been many different explicate orders, patterned on thoughts, actions and memories that differ from the world described by Newtonian physics. Here we can think about Heisenberg's claim that quantum theory could not be so crazy if an entire culture, that is, India, was based on similar premises. The unbroken and undivided totality, from which everything emerges, contains the potentialities of lived experience in the world. What follows develops the systemic implications of karmic entanglement for understanding the larger context surrounding George Floyd's murder.

Consciousness and property

Bohm's distinction between the fragmented modern explicate order and the larger unbroken implicate order shines a particular light on the tension between the sanctity of human life and the concept of the human as property, revealed in the protests following George Floyd's death. Indeed, the entitlement expressed in Derek Chauvin's treatment of Floyd as an object who could indefinitely be held down by his knee contrasts with Floyd's dying words: "I can't breathe". "I can't breathe" is a reference to the life force that runs through all sentient beings. "I can't breathe" is a warning that the life force is about to leave the body. The police officer's knee was held in place for nine minutes and twenty-nine seconds, which exceeds the amount of time a human can live without oxygen. Air and the ability to breathe are the most fundamental ingredients of life. In the absence of this vitalism, life expires. The thermodynamic decay of life into death is followed by new life. As life leaves one body, others are born. Birth is a manifestation of wave function collapse as the body is infused with the breath of life; death is the departure of the life force. The body is temporary, the subject of decay and the eventual cessation of life. 'Objective' material reality is not only changeable but continuously arising and declining in the context of a set of relational entanglements that extend across time and space, the relations-in-things.

In both Hinduism and Buddhism, the life force and intelligence that permeates the universe – or the implicate order, in Bohm's language – is referred to as prana. As Wendt (2015: 32) notes, the life force, or *élan vital* in the West, is an important element of quantum consciousness. His reference to humans as 'walking wave functions' reinforces the role of consciousness as a constituent of life (Wendt, 2015: 3). The Buddhist concept of self, as both real and non-real, thickens this relationship. The self has no intrinsic nature – it is not a thing – but is a product of the relationships that formed it. The idea that the sense of 'I', which all possess, reflects an unchanging essence rests on an illusion. The 'I' is instead an activity of 'I'-ing. The relationship between self and the various streams of consciousness is one of appropriating (*upadana)*, that is, the self is an activity of appropriating (Ganeri, 2012: 198). The self, in collapsing into consciousness of 'I', appropriates a space called 'mine' from which the self acts, navigates and directs the various streams of consciousness. This 'I' is shaped over time and changes within a relational web, within which meaning is super-positioned. Everything we see, hear and think is expressed and made manifest in the language available (the conventional apparatus) for seeing or perceiving in any particular context, and not least the language of 'I' and 'mine'. The self, in appropriating, draws on potentials stored in language, but also on a consciousness prior to language,[2] which holds the diversity of potentials and builds on habitual memory. The package of causes and conditions is like a bundle of disparate influences and streams of consciousness that have been woven together to make the particular 'I'. The bundle changes over time – or across time – as aggregates of consciousness and perception, relating to sight, hearing and feeling, engage with a relational world and are shaped by it. The qualities that shape the dependent arising of an individual in one lifetime carry over to another at the level of consciousness and affect. In Buddhism, the self is aggregated from streams of consciousness. The notion that the self is composed of properties that are unchanging, that is, an essence, is an illusion. The streams of consciousness are embodied but also entangled with a larger world.

The 'I' as an activity is holographic in so far as information that is generated in the whole is encoded in the parts. In Wendt's argument about quantum mind, entanglement is constituted through language. The thought that gives rise to the speech act, or what he refers to as 'wave function collapse into language', depends on external context and shared intersubjective meanings that are non-local and non-causal (Wendt, 2015: 255). Entanglement in language means that those who invoke it lose their identity as entirely separable. Yet separability or 'coherence', that is, the sense of 'I', are maintained, given that individuals are not entirely fused. The entanglement of parts and wholes is co-emergent (Wendt, 2015: 257). Co-emergence means that social structures actually are, physically, super-positions of shared mental states – social wave functions. The individual is a walking wave function that intra-acts with social functions (Wendt, 2015: 258). The structures are people and their practices, which are ontologically emergent from the entanglements of agents who constitute them. Buddhist karma or Bohm's implicate order suggest that the vast potentials of these social structures arise across space and time from the repetition and sedimentation of habitual patterns (see also Sheldrake, 2011). The patterns are recorded and encoded in memory that is written into the fabric of the universe. The latter will be further unpacked in later sections. The point here is that the relational self embodies the consciousness and relationships from which it has emerged, and cannot be entirely separated from these social structures. The entanglements are spatial in nature but also develop across time.

2 In Buddhism, this would refer to the *Alay-Vijnana*. Vijnana refers to awareness that connects, for example, the ear, a sense facility, to sound, a sense object. The Alay-Vijnana is the basis of all consciousness and contains impressions of all consciousness or seeds from which thoughts, desires, emotions and attachments grow. Karma is said to reside in a storehouse of consciousness.

Property and the appropriating I

In the fragmented explicate order, which is believed to consist of things, humans and objects might be said to occupy a level playing field. The fragmentation of life, from individuals and objects to cities, religions and political systems, remains intertwined within the whole implicate order, and it is the wholeness that is real. Human action is, however, guided by illusory perceptions that arise from fragmentary thought. The fragmentation is not a 'true copy of reality as it is' but a way of seeing (Bohm, [1980] 2002: 9). In this world, the chattel slave is conceivable. If humans and objects are like phenomena, then humans can be property just as objects are. Human owners and objects occupy the same world; in their separateness, both possess the properties of thing-ness, devoid of relationality, lacking in consciousness, emotion and compassion. An individual who possessed the latter qualities would find it inconceivable to own another human being and to treat them with the utmost brutality; and that brutality would not be enabled by law. Individual objects and property objects participate in the same world, which constitutes the enfolding and unfolding of a particular consciousness. Both exist in a separateness that rests on illusion.

A particular relational pattern based on an illusion of thing-ness runs throughout the parts and whole of the modern explicate order. The tendency to fragment is not specific to a particular time and place. Wang Chen, in the ancient classic *Dao of Peace*, identified the cause of suffering in this fragmentation, stating that 'as soon as things have names and people have emotions, love and hate arise and attack each other, warfare flourishes' (Sawyer, 1999: 18–9). But the Newtonian explicate order was distinct in its reification of thing-ness and exclusive focus on matter (Bohm, [1980] 2002: 5). Particular types of thing-ness and fragmentation emerged along with a colour line which specified types of property and provided ethical justifications for the appropriation of labour and the accumulation of profit.

Cheryl Harris (1993) provides an historical reflection on Whiteness as property in the US context. Her point is that Whiteness and property share the common premise of a right to exclude. White identity became the basis of racialized privilege that was ratified and legitimized in law. White was a type of status property (Harris, 1993: 1714) which evolved into new forms as White privilege became a legitimate and natural baseline. The 'cut' by which property status came to distinguish White and Black emerged over time. Racial lines among lower classes of Black and White people in the Americas were blurred until about 1800 (Roedinger, 1991: 24). However, from the seventeenth century on, the distinction between African and White indentured labour grew. Decreasing terms of service were introduced for White bond servants, and the increasing demand for labour led to a rapid increase in the number of Africans who were forcefully displaced to the colonies (Gossett, 1963: 30).

The construction of White identity and racial hierarchy went hand in hand with the expansion of a system of chattel slavery. The category of chattel slave was recognized in law by the 1660s, and between 1680 and 1682 the first slave codes appeared, codifying a deprivation of autonomy that was already in practice (Harris, 1993: 1718). The tension between the humanity of the enslaved and their status as property was evident in their categorization as 'three-fifths of all other person's' in the computation of votes for the House of Representatives, and in the use of Black women's bodies as a means to increase property (Burnham, 1987: 197–9). The property status of Whiteness went hand in hand with holding the enslaved as property. Whiteness was the property of free human beings; their enslaved property came to be defined by Blackness.

Harris demonstrates how divisions between the property of Whiteness and Black persons as property evolved and entered into law. The evolution only begs the question of how human property was justified in a country that was founded on a claim that all 'men' were created equal. Shotwell (2013: 318) notes that almost every ethical claim is based on a particular understanding of what humans are and what they ought to do to flourish. Likewise, every claim about whether an institution is good or bad relies on a set of assumptions about the kind of being an institution is good or bad for (Shotwell 2013; 319). The designation of the chattel slave as being only three-fifths human already qualified their standing. In any case, flourishing wasn't the objective but rather free labour and the political power of the enslavers. Karen Barad (2007: 384) states that ethics is about 'taking account of the entangled materializations of which we are part', and how particular designations contribute to the mattering of human beings. The mattering that arose from institutions of enslavement placed a high value on profit and thus property.

Max Weber's *Protestant Ethic and the Spirit of Capitalism* ([1904–5] 1958) is a useful point of departure for exploring the nature of being and institutions within capitalism. Weber provides an explanation for how a form of modern capitalism came about and how, once established, it was able to continue without attachment to its religious origins. He argues that Protestant Christianity, particularly in its Puritan or Calvinist forms, was a source of values and an orientation that resonated with the spirit of capitalism.[3] Particular historical circumstances in sixteenth- to mid-eighteenth-century Europe gave rise to entrepreneurial activity and 'reinvestment capitalism',

[3] Weber saw the influence of Puritanism on business in the US as a particularly strong exemplification of the thesis.

characteristic of the modern era. The continual reinvestment of profits into new enterprises was consistent with an ethic of ascetic Protestantism, in which hard work, profit and delayed gratification were valued.

Weber was not suggesting that greed or the desire to accumulate profit are exclusive to Enlightenment Europe or the US. What changed with the emergence of modern capitalism is the rational use of capital, the rational organization of labour as a calling and end in itself (Weber, [1904–5] 1958: 62) as well as the attachment of ethical meaning to the profit-making activity of the rational individual. The ethic suggests a change in the balance between production for need and production for profit, or 'the attainment of goods necessary to meet personal needs, [and] … a struggle for profit free from the limits set by needs' (Weber, [1904–5] 1958: 57). The subject of the *Protestant Ethic* is a disciplined rational actor whose primary concern is profit, who postpones enjoyment of the fruits of labour in order to reinvest and whose pursuit of profit is a calling.

While profit may require justification, profit does not by definition involve the enslavement of human beings. The colonies, and later the United States, offered possibilities to acquire property in the form of land from a vast uncultivated state of nature, which was there for the taking and increased the potential to generate individual wealth.[4] The philosopher John Locke, who was entangled in the social structure of profit and acquisition, provided a rationale for doing so. The individual on their own, in a state of nature, is limited in the degree to which they can generate wealth. In Locke's ([1690] 1978) state of nature, a place of complete freedom, it is difficult to accumulate surplus fruits. Apples go to waste because they are perishable ([1690] 1978: V: 139, 47). Surplus is the basis for profit. Locke's Labour theory of value asserts that property is a natural right derived from labour, along with life and liberty. Individual ownership is justified by the labour exerted to produce such goods (Locke, [1690] 1978: V: 130, 27).

The question is how one moves from a philosophy that values individual freedom and property, based on individual labour, to a justification for enslaving others. Within Locke's argument, in the state of nature every man has a property in his own person, and anyone who would attempt to take absolute power from them, making them a slave, puts himself in a state of war with him (Locke, [1690] 1978: V: 125). The Buddhist understanding of the self as an activity of appropriation shines a light on the logical conclusion

4 As Singer (1991: 102) stated, 'Property and sovereignty in the United States have a racial basis. The land was taken by force by white people from peoples of color thought by the conquerors to be racially inferior. The close relation of native peoples to the land was held to be no relation at all. To the conquerors, the land was "vacant." Yet it required trickery and force to wrest it from its occupants. This means that the title of every single parcel of property in the United States can be traced to a system of racial violence.'

of Locke's argument. The act of appropriating what is mine takes on a particular meaning within his Chapter 5 on property: what is mine extends to the appropriation of all that remains in a state of nature.

The act of appropriating mine extends the space 'owned' by the self. The state of nature, held in common and extending from 'the fruits of the earth, to the earth itself' (Locke, [1690] 1978: V: 132, 32), is appropriated and becomes property. The problem regards the limits of appropriation in the state of nature, and the command of God to humans, that is, the industrious and rational (Locke, [1690] 1978: V: 132, 34), who have been given dominion to subdue, till and cultivate such that nothing is spoiled, and the bounty of nature has a use for purposes of adding value and enhancing the exchange of the 'truly useful but perishable supports of life' (Locke, [1690] 1978: 139: 47). Not unlike the apple in the state of nature, property is appropriated and enclosed for private use. The invention of money introduces the potential for larger possession and a right to it (Locke, [1690] 1978: 134: 36). In the beginning it is the desire to have more than is needed that alters the intrinsic value of things, which arises from usefulness to human life (Locke, [1690] 1978: 134: 37). The true violation of God's will is not in the appropriation of 'mine' but in allowing products to spoil or go to waste. Those who remain in the state of nature, who are not putting the land to use, are distinguished from those commanded by God to till and cultivate.

Locke's *Second Treatise* influenced Thomas Jefferson in writing the Declaration of Rights. The legitimacy of government was to rest on natural rights, and citizens had a right to overthrow government if these rights weren't protected. Yet, in practice, some human beings were enslaved for purposes of generating profit for others, thereby losing any right to life, liberty, happiness or property themselves.[5] The contradiction did not go unnoticed, but was argued away based on claims that enslaved Africans remained in a state of nature and had not yet entered civil society (Lepore, 2018: 96). Because Africans remained in the state of nature they could be appropriated as property.

An ethical justification for acquiring profit within a capitalist system combined with arguments about the appropriation of nature. Black Africans were viewed as part of this nature, which justified the codification of a colour line which distinguished Whiteness as a property of status, and Blackness as the status of property, as expressed in the chattel slave. The justifications, legal categories and institutions that evolved around this illusion of humans as

[5] One might ask why the designers of the US Constitution chose to change Locke's 'life, liberty and property' to 'life, liberty and the pursuit of happiness'. Locke wrote that all individual are born with certain 'inalienable' natural rights, including to 'life, liberty and property', which means that they are God-given and can never be taken or even given away.

things illuminates Wendt's argument about entanglement in social structures. The entanglement was expressed in the thoughts and actions of large numbers of White Americans who contributed to the constitution of a distinction between those who could be free and those who would be property. The appropriation of 'I' and 'mine' took the form of owning other human beings within a particular explicate order.

The cycle of samsāra

The entangled social structures of relations-in-things could be given meaning in something like the popular notion of karma. Some Christians, although certainly not all, have argued that wealth is a sign of one's merit in the eyes of God, while poverty or enslavement are a form of punishment. Arguments of this kind emphasize an outcome as evidence of merit or demerit or of one's essential nature. The Buddhist concept of karma, which means action, is more concerned with the intention of action and its relational consequences. An act arising from the ego and the poison of greed generates a much different karma than an act of compassion that arises from awareness of relational entanglements with others. The nature of the action and its consequences are a function of the intentions of the appropriating 'I'.

The legal question in judging Derek Chauvin, the police officer who murdered George Floyd, was the degree to which his act was intentional. He was not charged with first-degree homicide, which suggests a clear intent to murder, but rather with second- and third-degree murder and second-degree manslaughter, which are less intentional. In contrast to the legal specifications, one might say that the intention to murder was expressed in the act of holding his knee on Floyd's neck for three times the three minutes that human life can be sustained without air. The act arose within a relational system where he, as a White police officer, expressed an entitlement to act as he did, in public and while being filmed, without serious repercussions. The conviction of a White police officer for the murder of a Black man was unprecedented in Minnesota. The conviction of Chauvin represented a shift of consciousness with potential reverberations into the future. What goes around comes around. The universe is one of movement, of a multiplicity of events and happenings, and as such, intention can at any one moment manifest itself in multiple and overlapping ways. Here we want to explore the intention to experience pleasure and avoid suffering, two intentions which, consistent with the Four Noble Truths of Buddhism, constitute karma as a habit of action which sediments through the repetition of particular patterns of relationship.

In developing a theory of karma, Framaran (2018) asks why it is that we return to particular types of action. We do so to repeat a pleasurable experience (I enjoyed eating chocolate cake and want to have the experience again) or to avoid an unpleasant experience (having become ill from eating chocolate cake, I will avoid it in the future). Desire and aversion can in this respect result in the repetition of action (karma) that becomes habitual (karmasaya, samskara)[6], that is, I repeat a particular form of action ad infinitum in order to experience pleasure or avoid pain. The action itself (for instance, eating cake) cannot be separated from the consequences of action (pleasure or illness) and the impact on further action in the future (a habit of enjoying or avoiding cake) and its implications (for example, weight gain or loss, with potential transgenerational consequences). The desire may simply persist until such time as it motivates action again, while meanwhile the actant is neither aware nor motivated by it (Framaran, 2018: 77). Desires motivate repetition of actions that produced pleasurable results in the past. Aversion predisposes one to avoid action that produced painful results in the past.

A notion of habits of desire and aversion seamlessly extends to institutions or cultures as well as individuals given that the 'I' is a product of dependent arising and is relational by nature. A desire for success expressed in the accumulation of wealth, and for freedom may go hand in hand with an aversion to poverty and persecution. Thinking about the habitual nature of karma takes us beyond entanglements through language, expressed in law, theory or ethics, to the emotions and memories generated through particular relational patterns that become systemic. Consider the dependent origination of the United States as a collective entity, shaped by incoming refugees or migrants of various kinds, initially from Europe. Aversion to an experience of poverty and/or persecution in the European context gave rise to mass emigration of Europeans to America. In the 'new' land, a desire to accumulate wealth went hand in hand with an aversion to repeating an earlier experience of persecution and poverty. In the developing capitalist economy, the pursuit of profit became encoded as a normative value, along with a focus on individualism and equality between individuals, all of which were reinforced by a Protestant ethic. The habit arose from a desire for individual wealth and freedom combined with an assumption of equality in their pursuit. Memory was the carrier of experience in one context, which informed desire and aversion, to another. From the rational argumentation regarding the justification for pursuing profit or enslaving human beings, we raise a question of the deeper emotional experience that would lead to the treatment of other humans as things rather than as relational beings.

6 Samskara doesn't have an exact parallel in English but is often translated as habit. It can refer to mental and emotional patterns, imprints or impressions left on the mind. It can be used to refer to anything that is conditioned or compounded by other things, which links it to the concept of dependent origination and the idea that one phenomenon will depend on conditions created by other phenomena. Samskaras are also the seeds that cause rebirth and the endless transmigration of samsāra.

Resmaa Menakem (2017), a trauma and somatic consultant in Minneapolis who has worked, among others, with the police, argues that the treatment of Black Americans, both historically and in the present, has its origins in White-on-White violence going back a thousand years to medieval Europe, well before the creation of the US.[7] His argument provides an opening for thinking about karma as part of a larger systemic relationship which develops from social habits, rather than as a divine punishment or reward meted out to individuals. The pattern that would later emerge in America was evident in the barbaric White-on-White practices of medieval Europe, and England in particular, as revealed in this brief snippet from the historian Barbara Tuchman (1978: 135):

> The tortures and punishments of civil justice customarily cut off hands and ears, racked, burned, flayed and pulled apart people's bodies. In everyday life, passers-by saw some criminal flogged with a knotted rope or chained upright in an iron collar. They passed corpses hanging on the gibbet and decapitated heads and quartered bodies impaled on stakes on the city walls.

Barbaric practices combined with the experience of the Great Plague (1665–66), which killed one hundred thousand people in London alone. English emigrants fled poverty, starvation and overcrowding, as well as imprisonment, torture and mutilation. Similar practices of punishment then became evident in the colonies. Somewhat later, suffering during the Highland Clearances in Scotland (1750–1860) and the Irish Famine (1845–52) gave rise to further waves of emigration to the US.

Menakem (2017: 59–61) argues that, until the seventeenth century, trauma was primarily inflicted on White bodies by other White bodies, and that indeed racialized references to White people and White body supremacy only appeared in the late seventeenth century, along with the laws discussed by Harris (1993). It is not that enslavement or the experience of trauma only emerged in medieval Europe; these can be seen across various cultures and civilizations, and, indeed, animals experience the excessive fear of trauma.[8] The new 'cut' transformed the dynamic as long-standing White-on-White practices were perpetrated against Blacks and Native Americans. White Europeans inflicted many of the same practices that they had fled against

---

7 His work builds on that of Joy DeGruy (2005), *The Post-Traumatic Slave Syndrome*, which explores in historical detail the violence and abuse inflicted on enslaved Africans in America and the continuing trauma inflicted on successor generations.

8 Although animals are able to 'shake off' their experience of extreme fear (Levine, 1997), and many earlier cultures had shaking rituals for trauma (Keeney, 2007), abilities that have been lost in modern culture.

enslaved Africans. Menakem (2017: 63) argues that 'to blow centuries of White-on-White trauma through millions of Black and red bodies' eased the conflict between powerful and less powerful Whites. Central to the reproduction of this relational pattern is what he refers to as 'dirty pain', which is pain arising from avoidance of grief, blaming others for past suffering and a denial of loss.

To return to the language of Buddhism, for both White and Black the seamlessness of dependent arising was broken, creating an illusion of individuals and objects. The movement of White peoples from one continent to another was an interruption of the causes and conditions from which self arises. The individual was disentangled from the collective to which he or she was born, thereby reinforcing his or her separateness, but then carried the formative causes and conditions, including emotions and actions associated with persecutors, to a new environment in the form of memory, where they became participants in creating a new entangled reality.

While both Europeans and Africans experienced an interruption and dislocation, the latter were not only forcefully displaced – that is, taken against their will and transported to a new land where they were cut off from the habits of their own dependent origination – they were also forced to take on habits that were imposed on them by others. Conformity was regulated by law, as was punishment for non-conformity. As chattel slaves, Africans became the property of White Europeans in the new land of freedom. As property, their free labour generated the wealth of free and equal individuals. The dislocation of forced displacement was not merely physical, relating to a loss of loved ones or physical damage to the body. The interruption also constituted a loss of the autonomy to live according to one's own habits of memory. The forced displacement into enslavement involved a double interruption: each individual was disentangled from the web of their dependent arising, but also from the habits of this entanglement. The physical dislocation to slavery, much like the process of colonization, involved what Ngugi (2009) refers to as 'dis-memberment', by which indigenous knowledge was supplanted and replaced by European memory.[9]

[9] The point is graphically and symbolically expressed by the European colonialist practice of cutting off the heads of African kings as carriers of knowledge and memory. The heads were then buried facing the earth, in opposition to indigenous burial practice (Ndlovu-Gatsheni, 2018: 11). The symbolic practice was bound up in the epistemicide of a continent, which involved not only forceful physical dislocation but dislocation of culture, tradition and knowledge systems as well. Ngugi (2012: 39) compares cultural dislocation to that of travel and the traveller's mind 'from the place he or she already knows to a foreign land. It is a process of continuous alienation from the base, a continuous process of looking at oneself from the outside of the self or with the lenses of a stranger.' The travel was not merely across space but also time.

My concern here is to highlight the historical entanglements, the emergence of the Black-White fault line, as a distinct relational 'cut', and the reproduction of a relational system which was propelled by the avoidance of persecution while inflicting very similar forms of persecution on others. The relational system has in part been sustained by silence and an inability to experience the pain of seeing the pattern for what it was and is. Maintaining an illusion of moral goodness requires looking away from the harm that has been done by the self, whether individual or collective. The result is a patterned collective experience, which is different on each side of the fault line. For instance, in the contemporary US, Menakem (2017: 27–8) further notes, White bodies tend to be seen as fragile and vulnerable, looking to police bodies for protection and safety; Black bodies are, paradoxically, seen as dangerously impervious to pain and in need of being controlled, yet also as a potential source of service and comfort to Whites.

> For most of our country's history, the Black body was forced to serve white bodies. It was seen as a tool to be purchased from slave traders; stacked on shelves in the bellies of slave ships; purchased at auction; made to plant, weed, and harvest crops; pressed into service in support of white families' comfort; and used to build a massive agricultural economy. (Menakem, 2017: 28)

In short, the Black body became an object of trade and profit, distinguished by its thing-ness, rather than a sentient human being who also had a right to the property and the promise of the new world. None of this was necessarily conscious, but all of it rested on illusion. The system produced a particular experience for Black Americans, which was reproduced in new ways with the end of formal enslavement. The relational system revolved around individuals, or those who could be free and equal, and the enslaved who were property and thus denied autonomy or property themselves.

As a construct, the Black-White fault line was larger than any one individual, but conditioned the behaviour of all within it. Any one individual is in and of themselves empty, that is, they are relational creatures who have arisen interdependently, shaped by family, institutions and culture, but also – of key significance here – by a historical and relational system driven by profit and potential greed, within which Black and White occupied different positions. The internal workings of any one individual mind will include the entanglements of language that arise from the causes and conditions that inform experience across time. However, the positionality and relationality of one time and space is changeable in another, as evident in the transition from White-on-White violence to White-on-Black violence. It is not a self with an essence that carries from one lifetime to another but aggregates of consciousness which are re-aggregated and repositioned in the next, perhaps allowing lessons from the past to present themselves as challenges in the future. Nor does suffering in one lifetime correlate with the good or evil essence of an individual in a prior life; samsāra, as a system, brings suffering to all who are caught up in it, if not always suffering of the same kind. An enslaver who may themselves have been enslaved embodied the earlier relational pattern but now shifts position from persecuted to perpetrator in order to avoid being a victim again. Indeed, the claim that settlers were slaves to the British, and would not be enslaved, even while owning slaves themselves, was central to the rhetoric of the American Revolutionary War (Dorsey, 2003). While resting on an illusion of separateness, whether as individual or thing, both positions are the product of a history of entanglement in relational patterns, whether carried in the DNA of individuals or in the stories, narratives and behaviours inherited from earlier generations.[10] Both were things woven through relations. Again, we can think of the contrasting yet entangled emotional habits expressed by Chauvin's act of entitlement and Floyd's words as he was dying, that is, "I can't breathe". The fruits of action are expressed within any one encounter, but they also potentially, as part of the experience of each side of the fault line, shape further encounters with others, which then sediment over time, becoming habitual and part of cultural and collective memory.

Those who identified themselves as citizens of the new land (who would not be slaves) generated profit through those who worked the land (and were enslaved). The idea that wealth is generated by the individual or a family that enslaves other humans rests on an illusion. The accumulation of wealth, and thus the aversion of poverty, requires that both the individual who profits, and the larger society of which they are part, sustain the illusion. The illusion sustains a moral order, a Protestant ethic and an economy based on exploitation of the labour of some for the profit of others. This is a relational order, within which the histories of different parties are entangled. Over time this dynamic develops into habitual modes of intra-action that reproduces karma of a particular kind.

The terms of the desire/aversion dynamic, that is, a desire to accumulate and an aversion to poverty or being enslaved, were dictated by the interests of those seeking profit, as well as by categories of ethics, economy and race. The enslaver-enslaved relationship and the violent habits that formed around it were wrapped in a veil of illusion and moral order. Practices such as policing arose from habits with their origins, among other places, in the capture of escaped slaves (Potter, 2013; Hansen, 2019), which mutated into new forms with the end of slavery. Likewise, habits such as lynching on the

[10] There is increasing evidence within the study of epigenetics of the transfer of trauma across generations.

part of vigilante groups, not least the Klu Klux Klan, as well as later forms of police brutality, kept aspects of the enslaver-enslaved relationship in place (McLaughlin, 2020; Pilkington, 2020). The Thirteenth Amendment, which contains the first mention of slavery in the US constitution, declared an end to slavery, except as punishment for crime, but provided the backdrop against which a disproportionate number of Black men have been arrested and held in prison, where they continue to work for free (Duvernay, 2016). Legal frameworks such as 'restricted covenants' severely constrained the property available for purchase by Black persons even in progressive northern cities such as Minneapolis (Delegard and Ehrman-Solberg, 2017), which presented a further obstacle to wealth creation and inheritance. Events such as the massacre in Tulsa, Oklahoma, in 1921 destroyed an emerging 'Black Wall Street', all to the end of keeping Black Americans in their place while also preventing the emergence of a level playing field in practice (Fain, 2017). The development of a particular desire-aversion dynamic, combined with an illusory notion of self as separate, reproduced and sustained a relational system that kept the karmic resonance of enslavement alive over time and across generations. The relational pattern began on another continent. The entanglements persisted long after the formal institutions of slavery had ended.

White-Black intra-action takes place within a context of pre-existing rules, both constitutive and regulative, that have become habitual, but which do not shape the behaviour of each party in the same way. There is a history of intra-action between Black men and the police that is deeply entrenched. As Black Lives Matter highlighted, the pattern has involved disproportionate punishment of Black men relative to their White counterparts (Ghandnoosh, 2015). Black parents have developed a habit of speaking with their children about how to engage with the police and what to do if apprehended while going about their daily lives. White parents don't generally have this conversation with their children. The history shapes and reinforces the engagement of each with the other. The intra-actions further sediment habits which are reinforced with each single act of violence. The individuals themselves are impacted over time by patterns of differentiation that carry resonances of these habits within them, and are predisposed or even hardwired to responses that reinforce the habit. Thoughts of, for instance, anger, hate or fear are not purely individual but are entangled with a habit of seeing that goes back generations. Sarah Ahmed (2004: 53–60) argues that these are manifestations of sedimented histories that remain open.[11] Emotional entanglements exist across time. A 'habit of seeing' becomes bound up in the

[11] Her reflection on Audre Lorde's encounter with a White woman on a train as a child is a powerful exploration in the context of a discussion of hate crime.

formulation of intent and action that reproduces the Black unseen, making it difficult for those positioned as hidden to move and reposition the self, to give it a place of belonging, long after the formal institutions of slavery have been abolished. The sedimented histories belong to a global context, of which the system of enslavement in the US was only a part.

Seen and unseen

According to Wittgenstein (1958: para. 129), we often do not see that which is overly familiar.[12] He was referring to everyday language, but a further distinction can be made between the visual stimulation of the eyes and the extent to which we actually see or don't see. On the one hand, we see the image of a human person before us. On the other hand, what we see is already heavily encoded in language by the time the person is placed in the category of, for instance, witch, terrorist, slave or king.[13] The meaning of a category is shaped by historical context; for example, the semantic and moral field surrounding a witch would have been much different in the sixteenth century than in the twenty-first, and the difference regards the extent to which the witch is considered to be 'real'. The sense of being seen or unseen raises a somewhat different, although related issue, regarding the extent to which the humanity of the subject is acknowledged. Certain categories, such as slaves, witches and terrorists, provide a measure of the humanity of the subject, making it possible to suspend certain assumptions about their moral status and thus shape action, including exclusion or violence toward them. Seeing the other as a slave, witch or terrorist pushes their humanity from view. Arguments that associate poverty with badness solidify a position in which those less fortunate are culpable for their own suffering.

Patterns of seeing and not-seeing also form the contours of memory within historical power relationships, or that memory which, according to Barad (2010: 260), is 'written into the fabric of the world'. The surface view is one of, for instance, enslavers who do not see the enslaved as humans and who thus are capable of the most atrocious acts towards them. This not-seeing relies on a social and legal context that denies the autonomy of certain others due to their being seen as less than human, and which legitimizes access to their bodies and punishment of them in often barbaric ways. There is a tendency to take language at face value, as a representation of reality 'as

[12] There is a noticeable resemblance between Wittgenstein's thought and both quantum physics and Buddhism. See, for instance, Stenholm (2011), Wienpahl (1979) and Anderson (1985).

[13] This has been an important theme in the literature on visual politics, as in, for instance, Bleiker (2018).

it is', rather than as a tool by which reality is created and navigated. The language at one and the same time makes it possible to see from a particular position in social space while also creating blind spots that keep suffering hidden from view.

Entanglement in the language and conversation surrounding slavery and the slave trade produced blind spots that limited the ability of even progressive and critical thinkers of the time to see the suffering that arose from these practices. David Hume's concept of 'commerce and conversation' reveals both the process and the problem. According to Kratochwil (2018: 354–6), the concept arose from a critique of the assumptions of enlightenment science regarding what it means to know, and rested on a claim that knowledge is produced in context and through communication. At the same time, 'conversation' within Hume's own context arose from racialized distinctions and from 'commerce' that commodified human beings. Hume himself had investments in slave plantations (Waldman, 2014). While he condemned Britain's empire of monopoly, war and conquest, as Ince (2018) argues, Hume ignored the relationship between transatlantic slavery and commerce, and confined his criticism of slavery to its ancient feudal or Asian manifestations. John Locke, whose ideas influenced the US social contract, was a major investor in the slave trade and participated in drafting the Fundamental Constitutions of Carolina, and Article 110 in particular, which gave enslavers absolute power over the enslaved (Uzgalis, 2018). A similar blind spot and contradiction is evident in Jefferson, one of the architects of the US Constitution, who believed that slavery should be abolished in America, yet who owned 600 people over the course of his lifetime (Smithsonian, 2019). Likewise, Abraham Lincoln, the author of the Emancipation Proclamation, demonstrated ambivalence in seeing slavery. His funeral address following the Civil War, as Meister (1999: 141) notes, provided a moral framework by which the defeated South could avoid the perception of humiliating punishment for slavery and secession and accept a Northern victory. While it is implied and assumed that slaves would be emancipated from their Southern owners, the humanity of the slaves and the continuing violation of their humanity were largely unseen as subjects of reconciliation. As Lepore (2018: 296) notes, Lincoln did not mention slavery in the address, although during his earlier election campaign he complained about the silence surrounding it:

> You must not say anything about it in the free states *because it is not here*. You must not say anything about it in the slave states *because it is there*. You must not say anything about it in the pulpit, because that is religion and has nothing to do with it. You must not say anything about it in politics *because that will disturb the security of 'my place'*. There

> is no place to talk about it as being a wrong, although you say yourself it *is* a wrong. (Quoted in Lepore, 2018: 279, emphasis in original)

Slavery was surrounded by silence on all sides, as if it was not there, even while there was pushback from abolitionists, both White and Black. The 'conversation' was between those whose power was bound up in commerce, with archives, diaries and ledgers controlled by slave owners and slave traders, within which the victims appear as only a name, as the property of someone, with nothing that would constitute their humanity. With abolition, the former enslavers were compensated for the loss of property rather than the human beings who had suffered enslavement. Hume, Locke, Jefferson and Lincoln were entangled in social structures, including an ethic that encouraged the accumulation of fruits, that is, the ego desire to pursue profit. The failure to see slavery may have been more or less intentional, or more bound up in justifications and theorizations of that order.[14] In any case, it would be difficult for anyone to engage in these activities while seeing and acknowledging the suffering they created. The two cannot be viewed simultaneously. What was, and arguably is still seen is Whiteness. As Illumine (2019) notes:

> The construct of race has been a defining feature of the modern world … a form of categorisation that has impacted on the entirety of the world's population through its distinctions between white and non-white peoples based upon their global origins. Despite the fact that it is a social construct lacking any form of genetic evidence, thoroughly refuted in biological terms, whiteness as a form of political dominance has been amazingly adept at retaining its power over the last four centuries.

Whiteness, as a racialized category, is on the one hand associated with the universal, and on the other provides an illusion of moral innocence while erasing the violence inflicted on the unseen.[15] Martin Berger (2005: 173–4), in an analysis of Whiteness in American visual culture, argues that race as it emerged in early modern Europe constructed an imagined boundary between Christian European and Jewish and African populations. From

---

[14] In her powerful argument regarding the erasure of race in Kant's thought, Jasmine Gani (2017) places Kant's universalizing theory in its historical context, arguing that it is the very universalism of his law of hospitality that erases the historical and racist context in which it was conceived.

[15] On the various ways that Whiteness is expressed, particularly within international relations theories, see, for example, Sabaratnam (2020), Krishna (2001), Henderson (2013) and Howell and Richter Montpetit (2020).

this boundary, those distant from Christian European norms were defined in terms of physical differences and subjected to a range of racially discriminatory attitudes and policies.

The face/vase picture created by the Danish psychologist Edgar Rubin presents the ambiguous figure of two white faces against a black background, which can also be inverted to a black vase against a white background. Both cannot, however, be seen at once (Goodale and Milner, 2013: 189). The original image had Black faces and a white vase; the version shown in Figure 3 highlights the extent to which a focus on Whiteness makes the black object unseen and unseeable in the absence of a reorientation. The White human subject is superimposed over the black material object.

While a function of habit, the relationship between that which is hidden and that which can be seen is not static but changeable. For instance, charging the White policeman in Minneapolis with murder went hand in hand with seeing the humanity not only of Floyd but of other victims of police violence. This changeability has several implications. First, if karma arises from an entangled relational consciousness, with traces or echoes of past life, rather than from an essential soul or self, then the positioning of rebirth within a relational order may change across lifetimes. Death and life are about the thermodynamic decay and reconstitution of matter and

Figure 3: Face/vase image, Edgar Rubin, Danish psychologist

Source: Adapted from Wikimedia Commons

consciousness in new bodies and life. Thus, one's position in the present on one side of the colour fault line may change in another lifetime, which creates the conditions for consciousness and seeing from a different position in space and time and thus for learning. The dissociation of rebirth from a particular self or soul suggests a potential re-aggregation of streams of consciousness activated by the intention-action-fruits of individuals on each side of the fault line.

Intention-action-fruits

The fault line is a product of a particular relational 'cut' (Barad, 2007) formed around Black and White. In the US context, any one body has been categorized as one or the other, despite a history of rape of Black enslaved women by their White enslavers which makes the binary nonsensical except as a category of experience. The binary arises from a historically specific dynamic that has sedimented and been reproduced in new forms over time. As relationships change, the relational self changes along with them (Shotwell, 2013: 322). The Buddhist theory of non-self places people in a dynamic encounter with one another and their environment, and acknowledges that the social process is continuously unfolding and along with it the way that people condition one another (Brazier, 2003: 138–9). Karmic residues from streams of consciousness may combine in new bundles as potentials collapse into human form. While Buddhism cannot comfortably be cast in the language of free will and determinism, as discussed in Snapshot 2, it does contain a potential for something like the former in so far as acknowledging the illusion, not least of self, can contribute to mindfulness of emotions and motives, thereby providing insight into how these and other factors impact on action. Indeed, the awakening of Buddha is about overcoming the illusion and turning away from the ego self to the relational self which, at the point of enlightenment, brings an end to the process of rebirth and the cycle of samsāra. While intention is not identical to free will, it is not rigidly determined (Harvey, 2007: 84). Mindfulness of relational entanglement can activate different intention-action-fruit potentials.

On the one hand, the three poisons, that is, illusion (ignorance), greed and hatred, condition a relational system that is formed around a fault line of Black and White with its origins in a system of enslavement. The illusion of individual labour propels commerce arising from profit, and potentially greed, which generates a relational dynamic that sediments in relations of hate, and related emotions, across the fault line, which continues to be reproduced in different forms long after the formal institutions of slavery have been abolished. The re-aggregation of consciousness across time also introduces further potentials for resistance and change. Action from an entangled position within a sedimented history of division between Black and White brings past action into the present where it becomes future. Reacting from within sedimented and entangled memory, and thus from within illusion, reinforces habits of ego and holds an imbalanced world in place. The wealthy White accumulate profit and wealth, buttressed by karmic justifications that wealth equals goodness and that individuals are responsible for their lot. The outcome thus implies the action. Wealth is a sign of goodness; I am wealthy, ergo I am good. The wealth-equals-goodness argument relies on an inversion of the Buddhist logic of karma. It begins with the fruits of ego, that is, profit and wealth (derived from enslaved labour), and uses these as evidence for goodness in this life or a past life.

On the other hand, karmic resonances of 'right' intention may collapse potentials into a new body on a different side of the fault line, such that the structure is not determined but is always changing. The Buddhist karmic equation, like quantum consciousness, places emphasis in the first instance on intention, from which action and consequences follow. The lack of concern for outcomes is in part due to the positioning of the actant, that is, as a consciously entangled self rather than an illusory ego self. However, given the non-linear understanding of time, outcomes may not yet be evident in any material sense beyond a relational orientation to others in the present. The fruits of action/karma (for instance, the activation of a particular relational mode, such as harm or compassion) can be distinguished from an outcome (for example, an accumulation of wealth).

Within the Buddhist logic, the inverted narrative 'wealth = goodness' becomes intention: the intention to seek profit activates desire and outcomes that express the fruits of that desire, that is wealth. The intention may facilitate harm to others, for example through enslaving or more modern manifestations of the habit, in order to maximize profit. The inability to see the suffering that arises from this relational system depends on hate to keep the illusion alive, to justify not seeing and to shift culpability away from the ego self. The intention and the action generate bad karma. The consequence of harm is a severed relationship, defined by a perpetrator and a victim. On the part of the enslaved, this pulses through a legacy of destroyed families, the experience of rape and of children born from this act of violence, poverty and constraints on establishing autonomy and wealth, all of which reverberate across generations.

The hauntingly beautiful novel *Homegoing* by Yaa Gyasi (2016) spins a story of two African sisters, one sold into slavery and one who became a slave trader's wife, showing how the consequence of this separation and the institutions by which it was constituted cascade across generations from the Gold Coast of Africa to plantations in Mississippi, from missionary schools in Ghana to dive bars in Harlem. The novel shows the relational consequences of the slave trade for these two sisters and the generations that would follow from them. While not her subject matter, the descendants of enslavers, and of White society more broadly, also experience negative impacts, albeit different in kind, related for instance to a need to hate or to avoid any kind of emotional world where, for example, grief, guilt or shame might be experienced. They may also be victims of the profit norm. Not all Whites, of course, became wealthy (see Fox Richardson, 2020). Which half of the separation one is born into is less the issue than the production of a relational world to which both sides are born and which is characterized by division and fragmentation and structurally driven by the three poisons.

Yet there is an opening and, as revealed in Gyasi's novel, a potential for change which also recurs and is a force of reform throughout this evolutionary history. As stated earlier, the primary issue is intention and action from whatever position, rather than one's position in society. Samsāra refers to the reproduction of a particular kind of world of circuitous change and the ongoing cycle of death and rebirth. One may be predisposed to reproduce the sedimented structure from a particular position in social space, but one is not determined in doing so. In this respect, karma in the Buddhist tradition is less attached to societal position than a relational orientation to intention and action, whatever one's position. One can see examples of wealthy businessmen such as Oskar Schindler, who in the context of Nazi Germany acted to save Jews at great potential cost to himself such that the relational intention-action-fruits were of greater concern than ego or profit. The lives of more than a thousand mostly Polish-Jewish refugees were saved during the Holocaust because Schindler employed them in his factories. The end of the Spielberg film *Schindler's List* shows the karmic effects of these acts with the multiplication of children, grandchildren and great-grandchildren of those who survived the death camps because of Schindler's efforts.

There is also space for mindful intention from the position of the enslaved. The person who is now referred to as Harriet Tubman had been enslaved (then named Araminta 'Minty' Ross) in the South. Against all odds, she escaped from her enslaver and was successful in making the journey north. Against even greater odds, she returned to the South to assist others to escape to freedom, initially to Philadelphia, and subsequently, following the passage of the Fugitive Slave Act in 1850, to Canada. It was less the outcome, that is, her personal freedom, that motivated her intention and action than her ability as a free woman to help others escape to freedom. Every time she returned to assist others, she took the risk of being captured and again enslaved. Her intention-action-fruits were greater than the self; that is, her freedom was inseparable from the freedom of others.[16] Following a blow to the head, in the context of punishment by her enslaver, she developed an

[16] It is also worth mentioning that she became a military leader along with the abolitionists during the US Civil War.

uncanny ability to see in much the same manner as the ancient rishi seers, and to become attuned to a larger universe, which helped her to navigate multiple journeys along the Underground Railroad. The nonviolence of the US civil rights movement, close to a century later, was similarly motivated by a concern to change the relational fruits of a system of enslavement, to cut through the prison of White and Black, in order to recognize a common inheritance of life. A relational system based on hate and fear cannot be transformed by more hate and fear – their opposite, love and compassion, are required.

Conclusion

The focus of this snapshot has been the paradoxical constitution of the chattel slave as both human and property, situated within an explicate order where humans in their separateness, and lack of relationality, are not qualitatively different than objects. In the American context, as Harris (1993) argues, Whiteness was a property of status, while Blackness was the mark of property. Black and White do not constitute the only fault line of identity in the United States, or the only relationship between unseen and seen. The black-white fault-line is, however, fundamental, given its roots in a system of enslavement. It could be analysed more broadly in relation to other entanglements, as these relate, for instance, to the dislocation of Native Americans to make way for the plantations; the relationship to the indentured servitude of some white immigrants, or the place of women in American society. Unfortunately, there is not the space here to look more broadly at these entanglements, although the final snapshot does identify a few further entangled 'knots'.

The illusion that the 'wealthy are good' and the 'poor deserve their lot' is a justification that rests on an aversion-desire dynamic. The dynamic is driven by an unwillingness or an inability to see our relational entanglement and the reality of suffering. The illusion of permanence that drives the individual ego and the importance of its desires generates the reproduction of suffering in the world and the cycle of samsāra. Similarly, in Bohm's quantum framework the fragmentation and brokenness of the explicate order, which maintains thing-ness, can be contrasted with the unbroken interconnectedness of the implicate order, even while the two are entangled.

The veil of illusion, that is, the belief that the wealthy are rewarded for good behaviour in the past and the poor punished for bad behaviour, shares a family resemblance with another concept, even while resting on opposite assumptions. John Rawls' (1971) 'veil of ignorance' is a model for detaching ourselves from our personal circumstances and stepping behind a veil where we don't know our particular identity or circumstances. This repositioning facilitates more objective consideration of the social contract and how societies might address the worst effects of inequality. Rawls' liberal conception of self assumes the separate individual who rationally determines their self-interest, detached from a context of birth. It assumes that an individual detached from context or identity will see more objectively. Rawls seeks to push beyond a position of privilege and devise more equal social structures. The danger is that the liberal grounding in individual, rational self-interest and the social contract makes it difficult to see beyond the illusion of self as essence, the habit of ego and, in regard to the social context in question, a tendency to see Whiteness. Indeed, rather than stepping behind a curtain, which blocks out the detail, we instead want to see through a veil of illusion to the significance of harmful relational patterns that have come to constitute life not only in the US but globally. The ego self of Buddhism or the thing-ness of explicate order are not exclusive to the US context, or the West or modernity; however, social structures of entanglement within this order reify and justify the 'appropriating I', not as an illusion but as a reflection of reality 'as it is'.

The metaphors of stepping behind or seeing through a veil are more significant than may seem to be the case on the surface. 'Stepping behind' suggests there is some more universal perspective from which reality can be objectively measured. 'Seeing through' suggests there is no magic behind the veil, no truth as such, aside from the truth of emptiness, from which we arise together from the formlessness of the unbroken implicate order. The contrast rests on two distinct concepts of reality and world. The idea of a universal position and Archimedian perspective relies on a materialist, Newtonian physics that there is a truth, which can be discovered in the world, a material background that constrains and organizes what can be. By contrast, the truth of emptiness and dependent origination or quantum entanglement means that there is no background reality to be found behind the screen; rather, life is a product of intention and potential embedded in a web of relationships, within which we participate as both weavers and woven.

When the intention expressed by Rawls' model is recast in terms of karmic entanglement, it looks quite different. The central problem of suffering, of which inequality is one manifestation, arises from attachment to the illusion that the self is separate and contains an essence, which gives rise to a need for self-protection and surplus profit. Self-protection relies on an aversion to or turning away from suffering within a desire-aversion dynamic which arises from the impermanence of life. Seeing this veil of illusion means acknowledging the emptiness of self that is a function of our dependent origination within a relational and entangled world. The desire-aversion dynamic, in the context of US race relations, has sedimented the entanglement of Black and White. The dynamic is driven by habits of ego and of seeing which hold a historical relationship of suffering in place.

George Floyd's murder was a moment of seeing the illusion of US equality for all, evoking the history of slavery and the transatlantic slave trade and revealing the vulnerability of Black life. Trump's election to the US presidency 2016 was also a moment of seeing, but from a much different position in social space. As Hochschild (2016) argued, Trump, 'the emotions candidate', acknowledged the suffering of working Americans, many of whom were descendants of European immigrants, who felt a loss of place as women and people of colour moved to the front of the line. Many of these supporters have not profited from the American dream to the extent that they feel entitled to, which adds an interesting and important dynamic. A frequent response to Black Lives Matter protests, 'Our lives matter too', while true, misses the point. While some European immigrants were indentured servants or even slaves, the relationship was temporary and not defined by skin colour. The example highlights the extent to which all are caught up in a dynamic where the value of property and profit comes head to head with the value of life.

The dynamic is most stark in the Black/White distinction by which the former were, within a system of enslavement, defined as chattel property and later faced with obstacles to acquiring property of their own. The tension and contradiction are captured not only in the image of Floyd's murder but in the downing of statues that commemorate slave traders, where the destruction of material property was directly counterposed with the devaluation of those who historically, as human property, were deprived of their humanity. As the American novelist, poet and activist James Baldwin (1985) said, 'History … is the present – we, with every breath we take, every move we make, are History – and what goes around comes around.' Baldwin does not here suggest that the experience of African Americans was a result of their 'bad karma'; rather the karmic relationship is systemic and potentially changes with 'every move we make'. Both George Floyd's murder and the conviction of his perpetrator in April 2021 were moves that potentially shifted the parameters of this system, making it possible to see the inhumanity at its core, and the humanity of its victims.

Endings/Beginnings

At Home in the Universe

Partha Dasgupta (2021), the economist who published the first full review of the economic importance of nature, stated that 'nature is our home'. Good economics, he argued, requires that we manage it better. Long-term prosperity requires an assessment of what nature is capable of supplying and how this relates to human need. Extant economic models tend to value human prosperity and the generation of wealth. 'Nature is a blindspot in economics that we ignore at our peril', the review submitted. If nature is home, then home has neither been valued nor seen. Always there lurking in the background, nature needs to be brought into sight in order to save it – and us – from ourselves. Far from being in balance, the house is burning down, and radical global changes to production, consumption, finance and education are urgently needed.

The issue is not merely nature but also the possibility of further pandemics. The ongoing destruction of nature is a fundamental driver of diseases that cross from wildlife into humans. Biodiversity experts have argued that even more deadly pandemics are likely unless the destruction is halted (Carrington, 2020). In a guest article for IPBES,[1] Settele et al (2020) stated that *we* are the problem. Pandemics are a direct consequence of human activity, and in particular of global financial and economic systems that value economic growth at any cost. 'Rampant deforestation, uncontrolled expansion of agriculture intensive farming, mining and infrastructure development, as well as the exploitation of wild species have created a "perfect storm" for the spillover of disease', they state. As the Buddha suggested, when the house is burning down, practice is more important than theorizing, for instance,

[1] IPBES stands for the Intergovernmental Science-Policy Platform on Biodiversity and Ecosystem Services, which is an independent intergovernmental body established by states to strengthen the science-policy interface for biodiversity and ecosystem services.

about whether we are capable of action. While theory and practice are never entirely separate, the question here, in conclusion, is how to put the fire out and to find a home in nature and the universe, and what this means for who 'we' are and what we do.

The turning of a blind eye to nature and the human role in creating the conditions for pandemics is only one side of the problem. The pandemic has shone a light on the organization of the global house. The 'structure' stands in stark contrast to the ontological parity associated with home in nature, as highlighted in Snapshot 3. The pandemic revealed the hidden underbelly of global inequality, which is not new but had been largely unseen, or attributed to a lack of development arising from more local conditions. The revelations were most shocking in the US and the UK, among the world's wealthiest economies and the global defenders of human rights. Western democracy, human rights and freedom represent positive values that have made both countries an attractive destination for people suffering from persecution or in search of better lives. These values are important and should not be dismissed, although some rethinking of the ontological assumptions surrounding them may be in order. Democracy, human rights and freedom underlie the values expressed through institutions and in public discourse, but as layers of illusion were peeled away, a stark and contrasting picture emerged, which like the yinyang symbol, revealed a black background behind a white presence, and two very different experiences of life. With the suffering on full display, the illusions could no longer be sustained. The virus revealed aspects of the unseen in Western societies which have underpinned a particular form of life and have become damaging to life more generally, including human life.

Duyvendak (2011) notes that feelings of home, including safety and belonging, have often been attached to a physical territory. Against the backdrop of globalization, goods and peoples were increasingly in motion. The large numbers of immigrants, as well as the advancement of women and people of colour within Western societies, gave rise to a nostalgia and romanticization of the nation state, resting on an older concept of home. The nostalgia became bound up in the politics of Far Right parties, which in the UK and US bled into the politics of Brexit and the election of Donald Trump to the presidency, respectively. One Brexit referendum poster showed the then UKIP leader Nigel Farage superimposed over a sea of darker refugee and migrant faces. The photo was purportedly taken in the Balkans, thus nowhere near the United Kingdom; it nonetheless communicated a potential invasion by outsiders. The message served the purpose of cementing racial and cultural difference that thickened the borders between those who populate home and those who threaten it, that is, those coming in from outside.

Globalization weakened the territorialization of home in the nation state. With increased mobility came thinner attachments and a sense of detachment from any particular place. Places and products, increasingly uniform and commoditized, became interchangeable, as diverse cultures merged into a unified space of capitalist appropriation and consumerism. As Hochschild (2016) shows, the desire to pull up the drawbridge and to, for instance, 'make America great again' was in part a response to the loss of home and place wrought by globalization and feelings of loss as those who previously were at the back of the line moved to the front. More tragic is the significant number of people across the world who lost their place and belonging, both inside the traditional nation state home and outside. By mid-2020, the number of forcefully displaced persons in the world passed 80 million (UNHCR, 2020); that is, persons without any kind of home. The majority of these (45.7 million) were internally displaced, within their own nation state, and 26 million were refugees.

Far from providing a model for what a more global home in nature might look like, a globalized consumer culture has contributed to the extremes of both human inequality and devastation of the environment. The creation of consumer desire, for fashion for example, had a number of consequences. Industries relocated from Western societies to developing countries in order to reduce costs and increase profit, often employing people at less than minimal wages, including child workers.[2] The need for cheap labour also drove human trafficking and modern-day slavery.[3] Keeping costs low meant seeking locations with lax environmental laws, with the consequence of increased destruction of the environment. A karmic system, entangled with the past and reproducing ontological disparity, that is, inequality, sustained an economic machine that depended on large numbers of low-paid workers for the profit of a few. How might a repositioning of the apparatus facilitate a change of direction?

The purpose of a concluding chapter is usually to draw together the threads of an argument so that the reader will hopefully be convinced. The task is somewhat more complex in this case, given the non-linear nature of the exercise and the attempt to construct a hologram in words. The conclusion as such is not singular or linear in its progression from A to B to C. Instead, a series of overlapping threads constitute the whole. Drawing together the threads will hopefully reveal a multi-dimensional image that connects

2 The 2015 report by the US Department of Labor found that eighteen countries don't meet International Labour Organization recommendations. A 2013 International Labour Office report found that more than 250 million children are in employment, primarily in the textile and garment industries, and six million are in forced labour (Moulds, 2021).

3 Forty million, three hundred thousand people are in modern-day slavery, including 24.9 million in forced labour and 15.4 million in forced marriage, which is 5.4 victims for every 1,000 people in the world. Women and girls are disproportionately affected (ILO 2017).

the various landscapes captured in the snapshots, and which also assists in rethinking what it might mean to be 'at home in the universe' given that many of the assumptions that have shaped modern life and science do not contribute to this end.

The first thread reweaves the science/culture relationship and two intersecting debates, between the potential for quantum social science, on the one hand, and the potential for global international relations, on the other. The second thread revisits the significance of positionality for understanding the researcher as entangled with the apparatus of measurement and the object of investigation. The third weaves through the various snapshots and the assemblage of self, mind, action, strategy, structure and time. The fourth thread returns to the individual/relational dichotomy to examine the theorization of relationality, particularly by Asian scholars of international relations, as it relates to both historical and contemporary practice. The final section zooms in on a further thread regarding the seen and unseen, which thickened in the move from one snapshot to another.

Science and culture

Dasgupta's claim that our 'home is in nature' suggests a need to bring the ontological parity of nature to an economic model that has prioritized human prosperity. 'Rational economic man' is the same everywhere, or so the theory goes. By contrast, ontological parity assumes diversity, and biodiversity in particular. Ontological parity is the principle that, whatever the differences, there are no degrees of being. Historically and in practice, there has been no ontological parity between humans and nature, or even among humans. One theme throughout the book has been the resonance between Newtonian materialism and an ontological emphasis on 'things', as well as determinism, and its relationship to a culture that has valued profit and acquisition over nature. Hobbes emphasis on escaping the state of nature, Locke's justification for the appropriation of nature and Weber's argument about the relationship between the Protestant ethic and capitalism reveal a culture that began in the West but which is now global. Theoretically and in practice, this model gives ontological primacy to humans over nature. Historically, it has given rise to the global appropriation of nature in the name of property, profit and wealth.

In social theory, the liberal model promotes freedom, democracy, individual rights, habeas corpus and the rule of law. In historical practice, the theory applied only to those who were believed to have escaped the state of nature, excluding women and non-Caucasians, who became the servants and slaves of a machine that generated wealth, the acquisition of things and profit for a small minority of the global population. It was not only the Black enslaved who were chattel, as discussed in Snapshot 6. Up until the mid-nineteenth century, women in Britain and the US were legally considered the chattel of their husbands based on a doctrine of coverture inherited from English common law, which meant that no female person had a legal identity (Allgor, 2012). Covered by her father's identity until marriage, any property owned by a woman would be transferred to the husband on the day of wedding. Married women had no legal right to do business, sign contracts or appear in court. Native Americans during the same period were deprived of their lands and faced forced removal westwards, becoming 'domestic dependent nations' with a relationship to the United States that was akin to that of a ward and his guardian. However, being a ward, a concept usually associated with preparing for independence, meant, with the Dawes Act (1887–1933), assimilation into mainstream US society, where they might be 'safely guided from the night of barbarism into the fair dawn of Christian civilization.' Citizenship was not granted to all Native Americans until 1924 (Boxer, 2009), less than one hundred years ago.

In the West, the circle of rights has expanded; the previously excluded have more potential than previously to prosper. Previous victims of imperialism, such as India and China, have become major economic powers in the global economy. But the global distribution concentrates wealth disproportionately in male hands. Women and people of colour continue to occupy disproportionately the ranks of underpaid 'essential workers'. The world's richest 1 per cent have more than twice as much wealth as 6.9 billion people. Men own 50 per cent more wealth than women, and the twenty-two richest men have more wealth than all of the women in Africa (Oxfam, 2021). The distribution does *not* reflect ontological parity but rather an extreme imbalance of wealth and opportunity.

Within the classical model, science is clearly distinguished from both culture and religion. Binaries coexist in a contradictory and hierarchical relationship. The parallel between quantum physics and ancient philosophies forces a rethinking of the either/or construction of science or culture, whereby science belongs to the West and culture to everyone else. The ancient seers knew things about 'reality' several millennia before they were discovered by contemporary physicists. There are numerous other examples of aspects of science arising from other cultures, such as the significant role of Islam in the development of mathematics, or technological advances in China, such as gunpowder, the compass and the printing press, which were appropriated and reinvented in the West (Hobson, 2004).

The 'rishis' and meditation, as well as Chinese and Indian Aruyvedic medicine, belong to culture. Culture contains properties that are considered to be contrary to those of science. Cultures, like art or religion, are particular; science is universal. Culture is about human relationships, their rituals, folklore, emotions and habits, and is thus the opposite of the hard rationality, materialism and objectivity of science. But the two can also be understood in their complementarity, as expressed, for instance, in the mutual constitution of Western culture and science, which appears a contradiction in light of the assumed universality of the later. Both have rested on similar foundations of materialism, determinism, mechanism and locality, concepts that have infused Western thought and practices. Western philosophies that have been critical of possessive individualism and capitalism have a long history, but have also often taken a back seat to the superiority of science, as reflected by the positivist versus reflectivist debate in international relations.

Global international relations seeks to expand the conceptual and geographical world beyond its origins in the US and UK in order to become a truly global discipline, where other subaltern and excluded voices matter, where it might be possible to look at the assumptions that underpin international life from another angle. The discussion has been bogged down by the inability to see the science/culture relationship in anything other than mutually exclusive terms. Science, as we know it, is a particular way of seeing that is embedded within global power relationships that shape whose knowledge is valued. One way out of the impasse might be to develop theories of international relations with different cultural characteristics. Another option would be to open up the conversation about the nature of science to the potential contribution of thought beyond the West, thereby pushing toward the universal, while recognizing what cultural diversity might contribute to a richer understanding of science. Eun Yong-Soo (2021) problematizes claims that international relations already enjoys theoretical diversity, arguing that the debate confuses qualitative diversity at an ontological level with quantitative diversity at a representational level, which reproduces a closed territoriality. He suggests that 'we – our bodies and emotions, our thoughts and minds – ought to be "active" in interweaving (entanglement) with a wide variety of knowledge-making agents and agency – which are in a process of becoming that have not yet actualised themselves in a specific form' (Eun, 2021: 21).

The parallel between quantum physics and the ancient Asian philosophies brings the science/culture relationship into view, providing a space for rethinking the significance of this entanglement. The notion that the scientist or any other observer cannot be separated from the apparatus of observation suggests that culture will always impact on what is seen. However, the parallel emphasizes uncertainty and motion as well as positionality, which suggests a greater potential fluidity than is often assumed. As an experiment in itself, this book has sought to explore the extent to which it is possible and beneficial to adopt an apparatus that is not of my own inheritance. The practice happens all the time, but from a different angle, given that diverse cultures have become embedded in a global capitalist economy, and many scholars from other parts of the world pursue their education and career in the West.

The pandemic presented something of a challenge to any rethinking of the science/culture relationship. The political world had for several years been shaped by culture wars and accusations of 'fake news', both by those who would freely circulate lies and those who sought to restore some method to the madness. An argument that what is seen depends on one's position, which is changeable, seems dangerous in a context where there is no common measure of truth. In the context of the pandemic, governments across the globe claimed that their decisions were 'guided by the science', and not least in those locations that were politically divided by 'fake news'.

Two months after the election of Joe Biden to the presidency, on the eve of his inauguration, many of those who voted for President Trump clung to false information, purveyed by him, that the election had been stolen, despite the failure of sixty-some court cases and claims by his own officials, and not least the US Attorney General, that the election had been won by Biden, fair and square. That so many Trump supporters would continue to resist wearing masks at a time when the death toll was reaching 500,000 was mind boggling, as were stories of patients in hospital dying from the virus who still refused to acknowledge that the pandemic was real. The psychological mechanism by which it becomes possible to deny a reality that is staring you in the face is not the issue here. From the perspective of experience, 'fake news' had become, in the context of the pandemic, a part of the American social fabric, with dangerous consequences, as witnessed on 6 January 2021. The Capitol Building was invaded by a mob of White men in the name of what was said to be a stolen election. 'Truth' is not imposed on reality but emerges from a relational fabric, even while it is conventional in nature and brings with it an element of illusion.

This is part of me and I am part of it

I am an observer trying to reposition the apparatus, from a particular location in time and space, with a set of experiences behind me, in order to potentially see something different about the nature of the reality exposed by the COVID-19 pandemic. Layers of 'normality' have already been stripped away to reveal aspects of life and suffering that we usually avoid looking at. My own past as a Wittgensteinian constructivist connects with aspects of quantum physics as well as Buddhism (see, for example, Wienhahl, 1979; Stenholm, 2011), but the parallel, as the apparatus, deepens insights into the nature of life in a quantum universe, and its relevance for human and other life, informed by millennia-old traditions. Stepping beyond my own intellectual inheritance provided the tools for engaging with self and world from a different angle in a context of pandemic where many assumptions about 'normality' had already been turned on their head.

'My' positionality is in many respects a function of the central problem it addresses, that is, action and strategy in the midst of uncertainty. My 'agency' as an actant cannot be separated from a larger environment, 'the very disposition of things' (Jullien, 1995: 13), which shapes me even as 'I' try to make sense of it. While the ground for this work was prepared over a period of close to a decade, the final form was shaped by circumstances that differ dramatically from those in which it was conceived. The original plan, after receiving the Leverhulme fellowship, was to spend nine months abroad at an institute for Asian studies, away from my home institution, where I might benefit from the perspective of researchers from around the world and a more extensive literature than otherwise would be available. For reasons beyond my control, these plans fell apart at the last minute. The rapid change of plans was nothing less than chaotic. The circumstances were further transformed by the emergence of a pandemic and multiple subsequent lockdowns that followed. Access to materials and engagement with other scholars was then further limited. It is difficult to judge the extent to which the quality has suffered or may have been enhanced as a result. The product is surely different because of the continuous upheaval. Conditions of lockdown did provide a more experiential taste of the 'emptiness' at the core of these philosophies. Less engagement with people, fewer distractions and more attention to a shared global experience of isolation had an impact on my engagement with the research material and my experience of the unusual circumstances. In so far as the ancient philosophies are concerned more with practice and experience than theory, there were some benefits to detachment from the usual academic push and pull.

The other obvious question about positionality is whether I should be writing this book at all. I am a theorist of international politics rather than quantum physics or Asian philosophy. The book is all of these, and none of these. As an American by birth, and a British citizen through naturalization, and having family ancestry that probably could, were records available, be traced back to the Vikings, I have dependently arisen from the context of global power over the last few centuries. In so far as positionality is about experience in the world, I have neither been a colonizer nor colonized, torturer nor tortured, an enslaver nor enslaved, nor, to my knowledge, were my ancestors. Nonetheless, this is part of me and I am part of it. I am a citizen of parts of the world that in the present and in the past have been seen to be a force for good in the world, but which are also responsible for practices that have been a source of significant harm to many. Placed within experiential relationships of power, which have patterned activity across the globe over time, the knee on the neck involves two actants – the one using the knee and the other who can't breathe. Changing the dynamic of the societal relationships and habits exposed by the pandemic necessarily requires a change of consciousness and action, that is, of 'I'-ing', on the part of parties who are responsible for harm as well as the harmed. There is no unencumbered positionality, absent some kind of historical or social clothing, but it is nonetheless useful to reposition the apparatus to look at self, both individual and collective from a different angle than is 'normally' the case. The quote from the *Rig Veda*, that truth exists but has different names, suggests that none of us see the same 'reality', but the motion inherent in the universe provides some scope for changing the position of observation.

Whether as a woman, an experience that is often laced with harm, or as a citizen of one portion of world that has held the levers of power for several centuries, I am a part of this and this is a part of me. The world is not static but in motion, with multiple traversing fault lines extending across time and contemporary space where my Scandinavian and Northern European ancestors immigrated to Minnesota, intruding on native American lands, perhaps witnessing or participating in massacre. In any case, the presence of European ancestors in the geographical location of the Americas cannot be separated from the social construction of laws that provided them, or at least the men, with a degree of freedom and opportunity that exceeded what they left behind. Freedom and wealth were intertwined within a system that enslaved or destroyed other human beings or discriminated against them in other ways. The practices were bound up in habit as well as in law and the generation of the country's wealth.

Half a lifetime ago, in a context of renewed fear of nuclear conflagration, I began my migration to Europe, or back to Europe, if placed in entanglement across time. It was the Second Cold War, and an American president was threatening nuclear demonstration shots over Europe. Movements that at the time pushed back against this potential, of which I was a part, were intertwined with the phenomenon itself. I could go on with examples from my past but 'I' am not the point. The point is instead that 'I' am part of a larger whole and a global equation that is badly out of balance; 'I' thus have an ethical responsibility from whatever position I occupy, and with whatever resources are available, to look at it from a different angle, to see through the illusions attached to 'normal' life, to begin the rotation from ego to relational 'I'. The pandemic started the repositioning. The parallel provided the apparatus for seeing. From the space of this emptiness, the action that presented itself to me was writing this book. The path has been full of twists and turns as well as contradictions. One obvious contradiction is the claim to an experience of 'emptiness' in lockdown that involves a proliferation of words. In this respect, the effort may reflect some of the more problematic aspects of the mindfulness debate presented in Snapshot 1, or it has taken me one step further in reimagining a relational self within the world. This brings us to the central thread that runs through the assemblage of mind, action and strategy.

Mind, action and strategy

Repositioning the apparatus at the parallel between quantum impermanence, complementarity and entanglement and Eastern mind, action and strategy involves moving away from a concept of agency defined by the binary oppositions between mind or body, free will or determinism, agency or structure, to understand action as an 'assemblage'. An assemblage, in Deleuze and Guattari's (1987) construction, suggests that entities are never fixed, pre-given or permanently stable in their ontological form or location. The self and other entities are continuously changing as movements, arising from processes of individuation and differentiation, both human and non-human, transform the relationship between parts and whole. Mind, action and strategy are the conscious elements of activation, which potentially ripple across space and time, but these movements are entangled in a larger whole, which also relies on conceptions of self, structure and time. The non-linear assemblage functions differently than the agency of the autonomous human. An assemblage is a sequence of parts. An activation of thought, speech or movement sets in motion a particular enfolding and unfolding that intersects with other assemblages, the interdependence of which then changes the whole. The potentiality of any one activation is already shaped by the motion of other parts in wholes that extend holographically across time and space. Each of the snapshots provide an angle on a particular part, which cumulatively point towards a universal woven in the threads that connect them, even while no one depiction can capture 'truth' in its entirety.

Snapshot 1 revealed a paradox and a process of decentring the self. On the one hand, the shift from mindlessness to mindfulness is conceptual. As just discussed, an illusory notion of 'I' and 'mine', assumed to exist as a separate essence, shifts to an awareness of 'I'-ing' as an activity embedded in a world of relationships. Having made this shift, assumptions of essence are replaced by awareness that the narrative potentials of self are multiple. The self as narration provides a smoother transition between the individual and collective 'I' than a model derived from Newtonian thing-ness. For the Buddha, as well as for Hobbes, narratives are continuously in motion, which is a source of suffering and e-motion in the world. Settling on a specific narrative, as suggested by Hobbes' social contract, vests an authority with the power to act, which then removes the problem of proliferating narratives to the international level. The transition from a state of nature to the social contract is not usually discussed in terms of narrative. A closer look at Hobbes' text brings the unceasing motion, propelled by narrative and emotion, into view. The ontological implications are contrary to the thing-ness of realist international relations and point to an 'entangled realism'.

The problem of narrative motion was revealed in the context of the UK decision to withdraw from the European Union. Their departure from the EU on 31 January 2021 widened fractures within the UK, opening the way for a potential constitutional crisis. The narratives of those who did not want to depart from the EU remained alive and gained impetus, not least in Scotland, where support for independence increased after the withdrawal. In Northern Ireland, practical problems with the delivery of goods from the British mainland quickly became a source of conflict with the EU, against a background of 'vaccine wars', while also lending momentum to a border poll about the future of the island of Ireland.

The Hobbesian social contract is about settling on a single narrative of authority; Buddhist detachment is about taking some distance from the proliferation of narratives and words, for purposes of detaching oneself from the influence of destructive entanglements and calming the mind. Detachment from the multiple narratives, allowing them to float freely, brings a recognition that none of them refer to reality 'as it is'. In the Brexit context, the UK as sovereign, Britain as the mainland (without Northern Ireland), the Great Britain of empire, the United Kingdom as a partner in Europe, the constituent parts of the UK that seek independence (Scotland and Wales) – these are all social and relational constructions. They have been real in shaping historical and present experience, which is continuously changing, but non-real in that none of them represent an essence. Indeed, the territory changes from one construction to the other. Detachment provides a different position from which to explore the relational consequences of any one path, an opportunity to reflect on historical patterns of practice that have become habitual, and to ask different questions about who is suffering.

Snapshot 2 explored the relationship between emptiness of mind (no-mind) and action through three interrelated paradoxes posed by the 'Buddhist warrior'. The Buddhist warrior is itself paradoxical as it expresses a form of action within war, that is, fighting with weapons, that is informed by a nonviolent tradition. Buddhism is, in the first instance, associated with health and well-being, that is, mindfulness or meditation, rather than war. Action from 'emptiness' of mind, in a context of war, combines with highly developed skill to facilitate an ability to respond effectively in a context that is rapidly unfolding. The further paradox regards the ethical justification for such action in light of the high priority given to nonviolence and compassion within Buddhism.

Emptiness, or no-mind, is a calming, a detachment from internal narratives of fear, anxiety or anger which so often cloud the mind, making it impossible to tune into the larger world outside the self, including the sounds of one's surroundings and those of a potential enemy. Trauma is the flip side of emptiness, a flooding of the mind with narratives and images of battle, of emotions of guilt or victimhood. Mindfulness in the US military context is about resilience and preparing the soldier to return to battle. The paradox of the Buddhist warrior raises questions about the shape of war and the extent to which the practice can be justified, including whether it is a source of psychological harm to both the warrior and his or her victims, and whether it is an act of defense or seeks exploitation and profit.

The two snapshots of section one drew out the parallel between quantum impermanence, arising from the consciousness/matter relationship and thermodynamic decay, on the one hand, and, on the other hand, the life and death cycle, which, in Buddhist thought, is a source of suffering that is not merely physical but existential. Thermodynamic decay is an experience of all life, as is the desire to cling to some notion of essence. Buddhism highlights the problem of impermanence, the suffering to which it gives rise and a potential path out of suffering. The most advanced forms of meditation differ across schools of Buddhism, but the practice suggests a detachment from the habits and entanglements of one's dependent origination, leading ultimately to an emptiness and calmness of mind which facilitates an identification with the suffering of all life.

A central question within the mindfulness debate is whether the cultural origins of mindfulness or mediation have been lost in translation with the introduction of 'secular' mindfulness to the individualist, capitalist culture of the West. Rather than transforming capitalist culture, making it kinder and less greedy, mindfulness, critics argue, potentially becomes a mechanism that maximizes the efficiency and productivity of workers and soldiers, along with the profits of global corporations. Self/no-self and mind/no-mind raise a larger question about the relationship between ontological understandings of self and mind and a context driven by the pursuit of profit, which is a theme that recurs in Sections Two and Three.

Section II brought the parallel between Niels Bohr's concept of complementarity and the Daoist yinyang to an analysis of the dynamic relationship between polar opposites and what it means to act in rapidly unfolding circumstances. The environment is a context of action, but it is also defined by nature. The concept of wuwei entangles action in a relational whole in which the human actant is a part of nature rather than separate and standing above. Movement is, on the one hand, individuated and differentiated in its uniqueness, and on the other, shaped by consciousness of deference to others who are also a part of nature. Action is actionless. The activation unfolds with minimal effort to maximum effect, precisely because it is attuned to the currents of the larger environment. Seeing beyond the self to movements in one's larger environment involves taking account not only of the visible yang but also the yin, or unseen and hidden dynamics. The polar opposites are the dynamic source of power and co-creativity. The question is not one of whether the agent acts with free will or is determined but rather one of skill in navigating the potentials of a context. During the Anthropocene, human artefacts have been destructive of nature rather than complementary, such that we face the potential loss of the habitat from which life has arisen. Wuwei repositions the self and its movements such that the uniqueness of the actant or its artefact expresses a deference to others and nature.

The concept of wuwei, which is developed in Snapshot 3, entangles action in a relational whole such that the human actant is a part of nature, bringing the strategic thought of Sun Tzu to an analysis of the 'war' against COVID-19. The highest value for Sun Tzu is to win a war without fighting a war. One of the many reasons for avoiding war is that fighting generates hate, which may make it difficult to restore balance to a system, and thus may have lasting reverberations. In any case, fighting a virus, the subject of this snapshot, did not involve military engagement, although the number of US and UK deaths after less than a year of pandemic exceeded the numbers of deaths in multiple wars. The threat was real. COVID-19 was a formless enemy, and Sun Tzu provides insight into a strategy for navigating a relationship defined by polar opposites and contradiction. Like wuwei, strategy requires attentiveness to one's context, being able to see the non-present dynamics of yin that at any moment may reverse themselves and become yang, the present. Polar opposites, when they reach one extreme, will revert to the other. The good general acts decisively, communicates clearly with her population and seeks to divert the virus while it is weak rather than attempting to hold it behind barriers which easily burst once the virus has become too strong. These qualities were absent in hyper-masculine male leaders who pretended the virus wasn't there, sent mixed messages and flouted the rules of their own medical experts. The result was a see-saw, back and forth between locking down too late, opening too early and then having to lockdown again, with an unacceptably high death toll as a result.

When the UK death toll reached 100,000, pundits asked why the country was doing so poorly relative to others. A frequent answer pointed to existing and increasing economic and racial inequality. The NHS and other relevant institutions were not prepared, having been weakened by a decade of Conservative austerity, which corresponded to an increasing disparity between rich and poor. Black and Asian minorities were more likely to suffer from health inequality than Whites, more likely to be on the frontline, where they had increased contact with the virus, and more likely to be living in dense urban conditions, poor housing and in multi-generational families. The disproportionate impact of the pandemic on these populations was an expression of an inequality of experience (House of Commons, 2020). On the same day that the UK hit the 100,000 mark, the newly elected US president, Joe Biden (2021), stated, as he announced his

Presidential Directive on Racial Inequality, that "all people are created equal, and should be treated equally in their lives". The fundamental imbalance in the treatment of human beings within wealthy industrialized societies – even before looking at the treatment of other societies or nature – was stark. The wealth of 660 US billionaires increased by almost 40 per cent during the pandemic (Collins, 2021), while unemployment, reliance on food banks and potential evictions skyrocketed. The problem of inequality is global, and is not defined entirely by a wealthy West and poor others outside. According to the 35th annual Forbes' list of the world's billionaires, the number of billionaires increased by 660 in 2021 to 2755, with 86 per cent having more wealth than in 2020.[4] Inequality is woven through the fabric of societies on all corners of the earth, and is not new.[5] The inequality builds on a history of global entanglement, snapshots of which are the focus of Section III.

Section III turns to the quantum concept of entanglement and its relational implications. Complementarity means that polar opposites interpenetrate one another such that they are mutually inclusive. Effective action or strategy works with contradiction rather than attempting to eliminate it. It is attentive to the dynamic relationship between the non-present background of the present. A concept of entanglement provides further insight into the relational significance of this infusion across space and time. Entanglement suggests that even at a distance, the movement of one part of an entangled whole will correspond to an opposite movement by its counterpart. When X goes up, Y goes down. The movement of each may be distinct, but they are intertwined, even non-locally. What does it mean to say that human social and power structures are entangled? The relationship is usually conceived in terms of the powerful imposing their will on the less powerful, to the end of controlling them. As recorded by Thucydides, the strong do what they will and the weak what they must. Section III explores how that which Karen Barad (2007) refers to as 'intra-action' might work in practice in carving out the boundaries of difference.

Snapshot 5 zoomed in on Gandhi's experiment in nonviolence, which arose from a different 'measurement' of India, as independent, in contrast to that of Queen Victoria's 'jewel in the crown'. Gandhi, as a strategist, was, like the quantum scientist, bound up with the apparatus of measurement, and he provides an illustration of the quantum notion, articulated by the physicist John Wheeler, that we are participants in creating the world. A more classical understanding of resistance as a disturbance to an otherwise

[4] According to the *Forbes* list, the US in 2020, before the pandemic, was somewhat ahead, while the *Hurun* Global Rich list from China, put China in front.

[5] Just under half of the worlds billionaires are from the US or China. Most countries have more billionaires than they did before the pandemic began in March 2020.

stable order, arising from a clumsy measurement, was contrasted with the activation and enactment of potentials. Gandhi's measurement took place against the backdrop of the British Empire and a larger international system defined by sovereign states. In this respect, the objective of independence was very much a function of that larger world and sat uneasily with the more positive relational underpinnings of Gandhi's strategy of nonviolence. The British Empire was entangled in a world of sovereign states. Gandhi sought to detach India from its entanglement in empire while keeping the relationship to Britain intact, and, in this respect, was either/and rather than either/or. However, as India became a sovereign state, further entanglements of violence between Muslims and Hindus rose to the surface and led to the creation of a separate Pakistan.

Gandhi's concept of desireless action emphasizes the importance of detachment from the 'fruits of action', or ends defined by ego. Desireless action links to a concept of karma. Karma means action, but does not refer to a singular movement in localized space by a discrete unit. Karma is an activation of mind and action that is entangled across time. Karma has a somewhat different meaning in Buddhism than it does in Hinduism, although the two traditions are historically intertwined and share many concepts. Snapshot 6 represents a further repositioning of the apparatus, bringing karmic entanglement to a reflection on the systemic nature of race relations in the United States. The analysis investigates the intra-actions that have constituted a Black-White fault line, going back centuries, which is woven into the fabric of the country. The fault line and its maintenance have been bound up in a tense and contradictory relationship between property and human life, as expressed in the erstwhile legal concept of the chattel slave. The notion of an 'appropriating I' in Buddhism shines a particular light on the appropriation of human beings as property from a Lockean state of nature.

The Black-White fault line has been intractable, embodying habits that have persisted while also undergoing transformation across generations and intersecting with other fault lines of race, gender and wealth. Part of its persistence arises from a habit of seeing 'Whiteness', while the violence against Black people remains an unseen background. The intersectionality suggests none of us have an essence; experience in the world has, however, formed bodies in distinct ways. The inequality of power and economy in both contexts has been the yin that supports the yang dynamic of profit and greed. The resurfacing of this hidden dynamic within the pandemic was a further expression of the reversal of opposites, a force that is akin to a physical law of nature, which highlights the importance of harmony and balance.

The death of George Floyd and the protests that followed brought the transatlantic slave trade and the history that it produced into view, where it could no longer be ignored. Months later, Kamala Harris was elected vice president of the United States, the first woman, and the first Black Asian woman, to hold the position. She, along with the president, shaped the most racially and gender diverse administration in American history. With the conviction of Derek Chauvin – the police officer who held his knee on Floyd's neck for nine minutes and twenty-nine seconds – for murder, the first conviction of its kind in Minnesota history, the circle came round, portending a potential for a more national shift in a system of policing with roots, among other places, in the slave patrols.

Houses and homes

The snapshots not only move through the different parts within the assemblage of action, they also move progressively outward from self and mind to body and environment, to entanglements in empire and across time. Here we bring the individual/relationality contrast back to the context of the global international relations debate as it relates to finding a 'home in the universe'. Discussions of grounding the individual more squarely in a social and relational world often gravitate to an either/or assumption; that is, if the parts don't come first, they will lose their distinctness and freedom to act. The evolution of human rights revolved around either Western civil rights, involving various expressions of freedom from political interference, or the Eastern emphasis on subsistence rights, or a right to food, healthcare and so forth (Vincent, 1988). The question is whether it is possible to reconceptualize the relationship between actants and their environment such that individuals and their relationality are understood as either/and rather than either/or.

The two snapshots of Daoism suggest that individuation, differentiation and action are fundamental to relationality. The uniqueness of the individuated is assumed, but creativity is shared with the context within which the actant, whether human or otherwise, is embedded. Within Daoism, action, although more constrained than assumed to be the case for the autonomous actor, is shaped by relationships to others within a whole. Deference and respect, particularly for difference, are the crucial ingredients in the aesthetic composition of home. We can start by imagining the difference between a family in which everyone does exactly as they please, regardless of the consequences for others, and a family in which everyone expresses their needs and wants while taking into account those of others.

A contrast between the response of South and East Asian countries, of whatever political constitution, and the US and European countries to the pandemic is relevant here. Many Asian countries were much more effective in keeping the virus at bay, while people of Asian and African descent were disproportionately victims in the US and UK. The political response in Asia ranged from the strict lockdown in China to the effective testing and tracing systems implemented in South Korea and Singapore, among others. Arguably, their greater success in minimizing the level of infection and subsequent death had as much or more to do with an acceptance of a practice of mask wearing rather than any particular organization of society. The act is not about protecting merely the self but also those who are potentially impacted by individual decisions that would otherwise contribute to the spread of the virus.

While Western populations did overall comply with more or less harsh government guidelines, those who resisted evoked a human right to freedom and clung to the right to decide individual movements, even in conditions where this could be harmful to family and compatriots. Trump election rallies involved huge crowds, very few of whom wore masks (in contrast to the Black Lives Matters protestors who generally did). The demand of resisters to be mask free, among other 'normal' freedoms, found expression in an attempt in the American state of Michigan to kidnap the governor, an action which was fuelled by conspiracy theories. Brexiteers communicated a similar message about the sovereignty of the UK and its right to make its own decisions outside of a partnership with the EU. At both levels, the tone was often one of 'No one is going to tell me to do something I do not wish to do', regardless of the consequences for others.

This demand takes on a particular meaning against the backdrop of over 125,000 COVID-19 deaths in the UK, and over 500,000 deaths in the US in the space of a year. Compare this with the number of deaths in Southeast and East Asia as of 13 December 2020: Twenty-nine deaths in Singapore, thirty-five in Vietnam, sixty in Thailand, 580 in South Korea, 2562 in Japan and 4756 in China (Statista, 2020). Even taking into account that the US and UK numbers are from February 2021, and in the midst of a second wave, there is simply no comparison. The difference does not come down to a distinction between democratic and authoritarian regimes, as the full range is represented in the Asian statistics. At the beginning of 2021, before the mass vaccination programs began in the US and UK, the differences were stark. As wealthy countries hoarded vaccines, many countries in Asia, as well as Africa and the Middle East, became increasingly vulnerable to the virus.

Scholars of globalization, such as Peter Singer (2004), argue that we can no longer afford to think in purely individualist terms. The planet is too interconnected. The individual/relational contrast provides two distinct frameworks for thinking about what it means to be human. The tendency in the West has been to focus on individualism and to assume that any notion of a whole, from the group to society to the world, is the sum of individuals. Western thinking is of course more diverse than this, but individualist assumptions underpin the dominant neoliberal or neo-realist models. Constructivists, post-structuralists, feminists, post-humanists and new materialists have questioned this individualist ontology, recognizing that it contributes to the creation of hierarchies of difference. Anyone who does not fit with the ideal of the rational individual becomes the excluded other. These constructions, it is argued, have contributed to the Orientalization of large sections of the world.

The point is not that we should throw out everything Western because it is now 'bad'. Individual human rights have provided both protection to individuals and an important point of departure for groups resisting oppression across the globe. Sovereignty provides protection and a basis for governing populations within specific territories. The point is to look squarely at the harm that has also been done through a history of global entanglement, which continues to be reproduced, if in new ways and different places, and to reimagine the relationship between ego self – at whatever level – and relational self. To do so it is necessary to reposition the apparatus to also bring other thought systems into the conversation, to conceptualize individuality and relationality as being intertwined and mutually inclusive rather than either/or, and to rethink what human rights or sovereignty mean if the 'I' is no longer an essence but a conscious activation of relational potentials.

The concept of relationality is more at home in Asia, although it does have a place in international relations more broadly (see, for example, Jackson and Nexon, 1999; Zalewski, 2019; Kurki, 2020), and a dialogue between East and West about the concept is underway (see Nordin et al, 2019). In conceptualizing relationality, Chinese scholars of international relations have tended to look more to Confucianism than Daoism, or to some mix of the two. The Chinese philosopher Zhao Tinyang (2006, 2009, 2021) brings the Confucian concept of *tianxia* ('all under heaven') to contemporary global politics. He argues that the Westphalian (Western) conceptualization of world politics is responsible for the mess we are in and that taming the current chaos requires a new non-Western framework of world politics. The concept of tianxia is not a literal representation of the world as 'all under heaven'; it instead represents a way to understand the world and its problems in a more holistic way in contrast to the national/individual emphasis of Westphalian statehood. Zhao argues that the construction of a more global entity would provide a remedy for chaos. He claims that out of the present chaos, a new material and normative order will become possible through the articulation of a new conceptual order, tianxia, which would allow for a truly global view of the world.

Zhao presents a relational view of international order that would replace the emphasis on sovereign states with a more relational whole. By contrast, Qin Yaqing's (2016, 2018) *Zhongyong* dialectics suggests a process from which something new might emerge from the dialectical tensions of the present, which might include the sovereign state-global relational order contrast.[6] Zhongyong dialectics share a family resemblance with the

[6] Qin is writing more in the Confucian tradition, although this has interfaces with Daoism, and not least the dialectic.

Hegelian or Marxist dialectic, but emphasize harmony and balance rather than struggle. Qin (2016: 41) argues that Zhongyong dialectics are about inclusivity, complementarity and harmony. Opposites are interdependent and complementary, 'co-evolving into a new synthesis through dynamic processes that continuously maintain, adjust, and manage complex and fluid human relations so as to reach the ideal state of harmony.' Opposites do not cancel each other out but are rather engaged in a continuous process of becoming, in a contradictory but non-conflictual relationship. The harmonious synthesis that emerges is a new form of life which contains elements of both poles, without being reduced to one or the other, much as with the birth of a new child (Qin, 2016: 41). Harmony is, in Qin's argument, a universal principle of order and a natural state of human and all relations. The harmonious synthesis has a parallel in quantum entanglement. As Zohar and Marshall (1994: 258) state, the many parts of a quantum universe 'are interwoven, their boundaries and identities overlap, and through their doing so new reality is created … when quantum system A unites with quantum system B to form quantum system C, C is *greater* than the sum of its parts. ... Quantum relationship is not a means to an end in itself.' Reality, in its multiple potentials, unfolds through a continuous process of becoming.

A relational ontology emphasizes context, which means that what 'reality' is depends on the perspective from which the world is viewed. Knowledge arises more from an experience of the world 'as ours' in the sense that it is shaped by histories, practices and habits (Qin, 2012). Shih et al (2019: 1) further define relationality as a 'process of mutual constitution that reproduces imagined resemblances, where actors self identify with others on the basis of a Wittgensteinian network of perceived and overlapping similarities.' Given existing histories, practices and habits, which both shape perceptions of similarity and difference and contribute to a process of mutual constitution, how might different parts within the world begin to engage in ways that move away from the burning house toward a way home?

L.H.M. Ling (2013) argues that the US/China relationship should be approached less in terms of alternative conceptions of order or threat than through the mutual implication of West and East. In her argument, the US is in China and China is in the US, meaning that they are intertwined in a relationship rather than totally separate, as is evident in the large number of Americans of Chinese heritage as well as the dependence of the US economy on China, among other things. From a Daoist perspective, the US and China are mutually implicated rather than mutually exclusive. The choice isn't either/or but necessarily looks to potentials that might, through their dialectical engagement, materialize. A reorientation of the apparatus provides a different angle on existing hybridity, which might propel global life toward more positive relational potentials. China's foreign policy is already hybrid in the sense that it is a modern nation state within the Westphalian model and is also influenced by its own millennia-long traditions, from Daoist military strategy to the influence of Confucian ideas on politics, such as tianxia (Wang, 2013, 2017). The debate about China's rise has revolved around a question of whether China represents an imminent threat or should be understood in terms of its more classical traditions which emphasize harmony. Shih and Yin (2013: 60) highlight a shared concern within this debate about how a rising China will treat others. From a perspective outside China, the answer is either/or, good or bad. However, when approached from the perspective of China's self-understanding looking out, its self-role conception is hybrid. China's understanding of core national interests, which echoes the realist concern, can coexist with its civilizational values, which emphasize harmony. Examples of both can be found in its history; that is, its self-conception is either/and rather than either/or.

As Shih and Yin (2013: 70) further state, China's obsession with core national interests and peaceful coexistence reflects a historical learning process since the Opium Wars, which engendered a consciousness of inferiority vis-à-vis the West. The memory of a century of imperialist invasions, which resulted in territorial concessions, made China hyper-sensitive to questions of territorial integrity. A repositioning of the apparatus might provide a better view of the historical practices and habits of contemporary states, that is, a slice of their dependent origination within entangled relationships that are global. The period of Western expansion and Newtonian science is a very short part of a much longer history which brought scientific progress and lifted portions of the world out of poverty. Less positively, it rested on a conception of human beings as fundamentally isolated from one another (Hobbes) and of nature as being there for human appropriation (Locke). Through most of that history, women and people of colour were considered a part of nature. The acquisition of property (the fruits of appropriation) was considered to be more important than the lives of those who remained in the state of nature and who could thus be property. These assumptions have contributed to habits and practices, more often inflicted outside the nation-home, but which are also present within it, that have been a source of harm to many. These habits and practices are a part of the repertoire of what contemporary states do.

In 2021, China was accused of genocide against Uyghur Muslims, including widespread rape, sterilization of women and forced labour in cotton fields. The intention is not to minimize the concern but to instead look at it in the context of a history of habit, practice and learning informed by an entangled history. Modern states stare at each other across fault lines that have constituted them as separate and sovereign, each desiring to impose an order of their own making while blaming others for the ills of the international system. Historically these international 'selves' are deeply entangled. The actors and their economies cannot be separated from a history and an international system that was produced from their engagement.

For purposes of illustration, consider a brief snippet of a historical global entanglement in the eighteenth and nineteenth centuries, which broadens out the perspectives offered in Snapshots 5 and 6. Prior to establishing a presence in China, the British East India Company engaged in trade with the former empire, exchanging silver for luxury goods, in particular for Chinese tea for the British market. The expansion of the British market for tea gave rise to increased sugar consumption, most of which was obtained from its colonies in the Caribbean, where it was produced by enslaved labour (Farooqui, 2016). Enslaved labour came from Africa through the long-established transatlantic slave trade. In the south of the United States, enslaved labour made an unprecedented level of cotton production possible, which underwent a boom with the invention of the cotton gin in 1793, resulting in an even cheaper American product. American cotton supplied raw cotton for the Lancashire textile factories in Britain. The mechanization of both spinning and weaving between 1815 and 1840 dramatically increased productivity, resulting in lower prices and Britain's ability to outcompete Indian textiles in the world market (Marks, 2015: 102–3).

With the loss of cheap silver after the US Revolutionary War, British opium exports from India to China increased steadily from the 1790s in order to finance the purchase of luxury goods from China (Farooqui, 2016). Within a few years of the British conquest of Bengal, which was completed in 1765, the British governor general of India had established an opium monopoly there and pushed its sale in China (Marks, 2015: 120). By the early nineteenth century, opium grown in India replaced silver as the commodity of exchange with China (Hervia, 2003). After the 1830s, silver for the purchase of opium flowed from China to India, which then made it possible to send large quantities of silver back to Britain. In Bengal, poor farmers were compelled to transform fields that had grown grain in order to plant poppies for opium export (Derk, 2012: 57). The export of opium to China created high levels of addiction and serious economic and social disruption. China's attempts to suppress the opium trade led to two Opium Wars, both involving Britain, resulting in peace negotiations and the Treaty of Nanking (29 August 1842), which, among other things, ceded Hong Kong to Britain.

This is a mere snippet of a complex history, but it highlights a global entanglement between China, India, the Caribbean, the American colonies (or later the US) and Africa, all within the British empire. The inflow of silver from the sale of opium underpinned the dominance of the British pound sterling until World War I and funded the colonial administration in India (Hervia, 2003). The history has reverberations into the present, not least as expressed by China's desire to leave its 'century of humiliation' behind while itself exercising a heavy hand in the affairs of Xinjiang, Tibet and Hong Kong in ways that conflict with its own relational discourse,[7] not least in terms of respecting diversity.

The ethical challenge is to stop thinking in terms of 'us' and 'them', the West and the rest, and to recognize the extent to which East and West or North or South are historically entangled. The history has relational consequences rather than being simply about 'events' in the past. The West placed itself above others through bloody practices of control over several hundred years. The practices reverberate through to the present. The point is not to suggest that enslaved labour in Xinjiang is okay because enslavement also happened in the US, or that the West is bad and now should succumb to a new model imposed by another part of the world. Instead, it is to ask what it would mean to acknowledge and grieve this past without turning it into a weapon and a new blame game that reproduces similar violent habits and patterns in new forms and new places. What would it mean to rethink the global home, such that its furniture, the human artefacts, are more attuned to the diversity and ontological parity of nature, indeed, more attuned to the potential for compassion, so that qualitatively different potentials might collapse into form? We cannot overturn the past through, for instance, revolution; the habits and practices that have emerged from it are a part of who 'we' have become. We can, however, look at the habits and practices and become more mindful of the appropriating self, whether individual or state, rather than assuming that the problem lies exclusively with others. Such a repositioning isn't to be found in blueprints and linear logics but in micro-engagements that begin an orthogonal rotation from the appropriating self to the relational self. As the house is burning down, there is every reason to find a different way forward that recognizes the destructive nature of many historical and present entanglements in order to 'ride the shi' (Voelkner, 2019) towards something new.

The Chinese-threat thesis, as well as discussions of global order on both sides, focuses on a question of *either* China *or* the US as global hegemons. But the multi-perspectivity and indeterminism of quantum physics, or of some of the ancient Asian traditions examined here, focus less on material power and more on consciousness and intention, and thus produce the possibility to rethink a global polity that builds on the positive potentials of both individuality and relationality, conceiving the two as intertwined and in balance. As Qin (2018) highlights, human relations matter for a deeper

7 See Fierke and Alfonso (2018) for an analysis of China's relational discourse.

understanding of social action. Relationships are necessary to international relations and, as Shi et al (2019) argue, need to transcend the familiar binaries of East and West, North and South, great and small powers, rationality and relationality, and to constitute a balance of relationships. In contrast to the emphasis of realism on states driven by self-help and engaged in a balance of power, a balance of relationships emphasizes how states are socially intra-related actors who necessarily pursue a strategy of self-restraint in order to participate in networks of stable and long-term relationships (Shi et al, 2019).[8] Abandoning the individual or the sovereign state is not the point, but rather understanding the parts as intertwined with a global whole that is in need of care and compassion.

Seeing the way home

The snapshots moved through philosophical concepts from Buddhism, Daoism, Hinduism and back to Buddhism, exploring how the parallels to quantum physics, that is, impermanence, complementarity and entanglement, manifest themselves in macroscopic life. The physicist David Bohm ([1980] 2002: x) compared a series of still photographs to the actuality of a speeding car to make the point that we tend to think of life in terms of a series of static images while the actual experience of movement has within it a sense of an unbroken, undivided process of flow. Bohm asked whether thought could only provide abstract and simplified 'snapshots' of reality, or whether it might go further to grasp the living movement that is sensed in actual experience. In the West, the idea that the thinker and the ego are completely independent of the reality that is the subject of thought is widely accepted, he claims (Bohm, [1980] 2002: xi). While ego is no less pervasive in daily life and practice in the East, there is more philosophical space to consider the relationship between the fragmentation of thought and a larger unbroken reality in its totality. Daoism, Buddhism and Hinduism are not the same. Each involves a somewhat different position from which to observe the phenomenon in question, that is, what it means to act in an uncertain world and to be at 'home' in the universe. Together the snapshots provide glimpses of truth that are packaged in different conventional forms.

Life is temporary. It is both birth and death. Grieving the latter is no less important than celebrating the former. It is only in the one that the other is seen. The complementarity of life and death, good and evil, beauty and ugliness is a creative dynamic. Complementarity is the mutual implication of polar opposites in motion. While seeking balance, too great an imbalance

[8] Brent Steele (2019) makes a compatible and sustained argument that habits of restraint have a long-standing role in international relations.

will be followed by reversal to the opposite extreme. Reversals in the environment, the economy, politics and the pandemic happened and will continue to happen until some kind of harmony and balance is found. Navigating the uncertainty requires an ability to see both sides of a dialectical reality, including the hidden, and not only the one-dimensional image of good that states generally attach to themselves, which is a product of illusion. Historical entanglements are among the hidden and are fundamental to how relationality on the planet has evolved, including the habits and practices Qin refers to and the dynamics of self and other, power and wealth, that have added fuel to a house that is rapidly burning.

The latter points to a final thread, regarding the seen and unseen, which thickened from one snapshot to the next. The first snapshot of Buddhism, and of a cosmology of impermanence shared with quantum physics, began with the light shone by the pandemic on the inequality of suffering, both within societies and across the world. A concept of seeing as knowing, which goes back to the Hindu rishis and practices of meditation, was in the second snapshot brought to emptiness and no-mind in the context of battle. The snapshots of Daoism zoomed in on yinyang and the parallel to quantum complementarity as a dynamic relationship between seen and unseen, present and non-present. The seen/unseen dialectic is fundamental to the process of differentiation within nature. It is also a relationship to be conscious of and to navigate as an actant and strategist who with minimal effort and maximum effectiveness achieves their objectives. The final two snapshots revealed the role of the seen and unseen as binary oppositions in reproducing or transforming systemic power relationships and inequality. While the appearance of structure is one of stasis, maintaining the illusion requires the formation of habits that are not entirely stable. The unseen do not remain hidden entirely. They emerge into the present to be seen, for instance through practices of resistance, which may transform the context without entirely disentangling the fault lines upon which it rests. Habitual patterns of practice continue to be reproduced through seemingly new human artefacts, but the material 'reality' is continuously moulded by transformations of consciousness from past life, which propel movement forward.

The important point is that none of this is either/or. All of it is either/and. Once recognized, the human relationship to its own artefacts potentially changes. The tendency, going back to Hobbes, to ground human nature in thing-ness, attached to a mechanistic materialism, makes humans into 'self-maintaining engines of identical design, who cannot have other regarding aims' (Slomp, 2018: 113). Once the individual and all life are observed from the perspective of their relationality and the often hidden but constitutive dynamic of relational becoming, there is space for consciousness regarding our role as participants in starting the fire that is rapidly engulfing the house, as well as regarding our responsibility to nature and each other to construct a proper home, characterized not by mechanistic materialism but by care and nurture of that which 'we' hold in common, by an ontological parity in which every part has a place.

Faced with an onslaught of uncertainty imposed by a global experience of pandemic, and shaken by the exposure of layer upon layer of illusion peeled away in the process, the snapshots reflect on a range of contexts of uncertainty in which rapid action was required. The house is burning down but we want to return to 'normal' as quickly as possible. In January–February 2021, as US and UK deaths reached a milestone – or as one commentator said, a 'tombstone' – a further reversal took place. A new American president rejoined the international community, and not least the Paris Climate Accords and the World Health Organization. After a stumbling management of the pandemic itself, the UK was the first to roll out a vaccine which provided hope for the future. On the one hand, the rapid distribution of the vaccine gave rise to predictions that the pandemic, like the Spanish Flu pandemic of 1918, would be followed by another 'roaring twenties', as pent up energy following extended lockdown was released, but with little reflection on the further chronology of depression and world war that followed.

On the other hand, recognition of the milestones saw political leaders and populations turn to the grief surrounding such monumental suffering and death. While being 'guided by the science' had been a frequent mantra of the UK government, which at one point was responsible for the highest per capita death figures in the world, the new year of 2021 also brought a message from those suffering loss: the people who died aren't numbers. Each of one of them is part of a network of relationships extending outward from family and friends. The nightly recitation of numbers of infected, hospitalized and dead, while useful for gauging the progress of the virus or the battle against it, had the impact of dehumanizing those who were its victim. Grief cannot and should not be quantified. It is not a thing but instead must be felt, experienced and seen. One person who was interviewed stated that "it felt like the whole world was grieving". It is crucial to grieve, and to step into the 'emptiness' left by the pandemic, rather than rushing back to 'normal'. A more mindful global 'us' is needed in order to successfully navigate whatever lies ahead and to steer a course toward 'home in the universe'.

References

Abrahamson, Rita (2016) 'Assemblages', in Xavier Guillaume and Pinar Bilgin (eds) *Routledge Handbook of International Political Sociology*. London: Routledge.

Abrahamson, Rita and Michael C. Williams (2011) *Security Beyond the State: Private Security in International Politics*. New York: Cambridge University Press.

Acharya, Amitav (2014) 'Global International Relations (IR) and Regional Worlds: A New Agenda for International Studies', *International Studies Quarterly*, 58(4): 647–59.

Acharya, Amitav and Barry Buzan (2010) 'Why is there no Non-Western International Relations Theory? An Introduction', *International Relations of the Asia Pacific*, 7(3): 285–312.

Adam, Martin T. (2006) 'Nonviolence and Emptiness: Buddha, Gandhi and the Essence of Religion', *The Journal of the Faculty of Religious Studies*, 34: 1–14.

Adger, W. Neil, Jon Barnett, F.S. Chapin III and Heidi Ellemor (2011) 'This Must be the Place: Underrepresentation of Identity and Meaning in Climate Change Decisionmaking', *Global Environmental Politics*, 11(20): 1–21.

Ahmed, Sarah (2004) *The Cultural Politics of Emotions*. Edinburgh: University of Edinburgh Press.

Allan, Bentley (2018a) *Scientific Cosmology and International Orders*. Cambridge: Cambridge University Press.

Allan, Bentley (2018b) 'Social Action in Quantum Social Science', *Millennium*, 47(10): 87–98.

Allgor, Catherine (2012) 'Coverture: The Word You Probably Don't Know But Should', *National Women's History Museum*, [online] 4 September, available online from: https://www.womenshistory.org/articles/coverture-word-you-probably-dont-know-should

Allison, Elizabeth A. (2015) 'The Spiritual significance of Glaciers in an Age of Climate Change', *WIREs Climate Change*, 6: 493–508.

American Museum of Natural History (AMNH) (2021) 'The Great Debate: Quantum Theory', *American Museum of Natural* History, available online from: www.amnh.org/exhibitions/einstein/legacy/quantum-theory

Ames, Roger T. (1981) 'Wu-wei in "The Art of Rulership" Chapter of Huai Nan Tzu: Its Sources and Philosophical Orientation', *Philosophy East and West*, 31(2): 193–213.

Ames, Roger T. (1986) 'Taoism and the Nature of Nature', *Environmental Ethics*, 8: 317–49.

Ames, Roger T. (1993) *The Art of Warfare*, New York: Random House.

Ames, Roger T. and David Hall (2003) *Dao de Jing: 'Making This Life Significant'*. New York: Ballantine Books.

Anderson, Tyson (1985) 'Wittgenstein and Nāgārjuna's Paradox', *Philosophy East and West*, 35(2): 157–69.

Anievas, Alexander, Nivi Manchanda and Robbie Shilliam (2015) *Race and Racism in International Relations: Confronting the Global Colour Line.* Oxfordshire: Routledge.

Aradau, Claudia (2010) 'Security that Matters: Critical Infrastructure and Objects of Protection', *Security Dialogue*, 41(5): 491–514.

Arday, Jason and Heidi Sofia Mirza (2018) *Dismantling Race in Higher Education: Racism, Whiteness and Decolonising the Academy.* London: Palgrave.

Arfi, Badredine (2018) 'Challenges to a Quantum-Theoretic Social Theory', *Millennium*, 47(1): 99–113.

Baer, Ruth A. (2003) 'Mindfulness training as a clinical intervention: A conceptual and empirical review', *Clinical Psychology: Science and Practice*, 10(2): 125–43.

Baldwin, James (1985) *The Evidence of Things Not Seen.* New York: Henry Holt & Co.

Ball, Philip (2018) *Beyond Weird: Why Everything you thought you knew about Quantum Physics is … Different.* London: The Bodley Head.

Barad, Karen (2007) *Meeting the Universe Halfway: Quantum Physics and the Entanglement of Matter and Meaning.* Durham: Duke University Press.

Barad, Karen (2010)'Quantum Entanglements and Hauntological Relations of Inheritance: Dis/continuities, Space Time Enfoldings, and Justice-to-Come', *Derrida Today*, 3(2): 240–68.

Barnett, Jon (2000) 'Destabilising the Environment-Conflict Thesis', *Review of International Studies*, 26: 271–88.

Barnhart, Michael G. (1994) 'Sunyata, Textualism and Incommensurability', *Philosophy East and West*,44(4): 647–58.

Batchelor, Stephen (1998) *Buddhism Without Beliefs.* London: Bloomsbury.

Bauer, Alain (2007) *Isaac Newton's Freemasonry: The Alchemy of Science and Mysticism.* Rochester: Inner Traditions.

BBC News (2021) 'Covid-19: Keir Starmer on 100,000 deaths in the UK', [online] 26 January, available online from: https://www.bbc.com/news/av/uk-politics-55818127

Bell, Duncan and Srdjan Vucetic (2019) 'Brexit, CANZUK, and the Legacy of Empire', *The British Journal of Politics and International Relations*, 21(2): 367–82.

Benesch, Oleg (2014) *Inventing the Way of the Samurai: Nationalism, Internationalism and Bushido in Modern Japan.* Oxford: Oxford University Press.

Bennett, Jane (2010) 'The Agency of Assemblages and the North American Blackout', *Public Culture*, 17(3): 445–65.

Berenskötter, Felix (2014) 'Parameters of a National Biography', *European Journal of International Relations*, 20(1): 262–88.

Berger, Martin (2005) *Sight Unseen: Whiteness and American Visual Culture.* Berkeley: University of California Press.

Bergson, Henri (1911) *Matter and Memory.* Crows Nest, AU: George Allen and Unwin.

Bhambra, Gurminder K., Dalia Gebrial, Kerem Nişancıoğlu (eds) (2018) *Decolonising the University*. London: Pluto Press.

Bhandar, Brenna (2018) *Colonial Lives of Property: Law, Land and Racial Regimes of Ownership.* Durham: Durham University Press.

Bially Mattern, Janice (2005) *Ordering International Politics: Order, Identity and Representational Crisis.* Oxfordshire: Routledge.

Biden, Joseph (2021) 'Remarks by President Biden at Signing of an Executive Order on Racial Equality', *WhiteHouse.gov*, [online] 26 January, available online from: www.whitehouse.gov/briefing-room/presidential-actions/2021/01/20/executive-order-advancing-racial-equity-and-support-for-underserved-communities-through-the-federal-government/

Bilgin, Pinar (2010) 'The "Western Centrism" of Security Studies: "Blind Spot" or Constitutive Practice? Special Section on the Evolution of International Security Studies', *Security Dialogue*, 41(6): 615–22.

Bilgin, Pinar (2016) *The International in Security, Security in the International.* London: Routledge.

Bilgrami, Akeel (2008) 'Gandhi, Newton and the Enlightenment', *Philosophical Exchange*, 38(1): 2–18, available online from: https://core.ac.uk/download/pdf/233570732.pdf

Bilmes, Linda (2021) 'The Long-Term Costs of United States Care for Veterans of the Afghanistan and Iraq Wars', 18 August, *20 Years of War*, a Costs of War research series, Watson Institute, Brown University, available online from: https://watson.brown.edu/costsofwar/papers/2021/CareforVeterans

Bitbol, Michel (2018) 'Mathematical Demonstration and Experimental Activity: A Wittgensteinian Philosophy of Physics', *Philosophical Investigations*, 41(2): 188–203, available online from: https://philpapers.org/rec/BITMDA

Bleiker, Roland (1993) 'Neorealist Claims in Light of Ancient Chinese Philosophy: The Cultural Dimension of International Theory', *Millennium*, 22(3): 401–21.

Bleiker, Roland (ed) (2009) 'Editor's Introduction', *Alternatives*, 25(3): 271–2.

Bleiker, Roland (ed) (2018) *Visual Global Politics*. London: Routledge.

Bobbio, Norberto (1993) *Thomas Hobbes and the Natural Law Tradition*, trans. Daniela Gobetti. Chicago: University of Chicago Press.

Bohm, David ([1980] 2002) *Wholeness and the Implicate Order*. London: Routledge.

Bohr, Niels ([1961] 2010). *Atomic Physics and Human Knowledge*. New York: Dover Publications.

Boissoneault, Lorraine (2017) 'The Real Story of the Koh-in-Noor Diamond – And Why the British Won't Give it Back', *Smithsonian Magazine*, [online] 30 August, available online from: https://www.smithsonianmag.com/history/true-story-koh-i-noor-diamondand-why-british-wont-give-it-back-180964660/

Booth, Ken and Nicholas Wheeler (2009) *The Security Dilemma: Fear, Cooperation and Trust in World Politics*. London: Palgrave.

Boxer, Andrew (2009) 'Native Americans and the Federal Government', *History Review*, 64, available online from: www.historytoday.com/archive/native-americans-and-federal-government

Bowman, Norah (2019) 'Here/There/Everywhere: Quantum Models for Decolonizing Canadian State Onto-Epistemology', *Foundations of Science*, 26(1): 171–86.

Braidoitti, Rosi (2013) *The Posthuman*. Cambridge: Polity Press.

Brazier, Caroline (2003) *Buddhist Psychology*. London: Robinson.

Briggs, Morgan and Roland Bleiker (2010) 'Autoethnographic International Relations: Exploring the Self as a Source of Knowledge', *Review of International Studies*, 36(3): 779–98.

Bueger, Christian (2018) 'Territory, Authority, Expertise: Global Governance and the Counter-Piracy Assemblage', *European Journal of International Relations*, 24(3): 614–37.

Burgess, James Peter (2015) 'An Ethics of Security', in Gabi Schlag, Julian Junk and Christopher Daase (eds) *Transformations of Security Studies: Dialogues, Diversity and Discipline*. Oxfordshire: Routledge, pp 94–108.

Burgess, James Peter (2018) 'Science Blurring its Edges into Spirit: The Quantum Path to Atma', *Millennium*, 47(1): 128–41.

Burnham, Margaret (1987) 'An Impossible Marriage: Slave Law and Family Law', *Law & Inequality187*, 5(2): 187–225.

Byg, Anja and Jan Salick (2009) 'Local Perspectives on a Global Phenomenon – Climate Change in Eastern Tibetan Villages', *Global Environmental Change*, 19: 156–66.

Callahan, William and Elena Barantbantseva (2012) *China Orders the World: Normative Soft Power and Foreign Policy*, Baltimore: Johns Hopkins University Press.

Capra, Fritjof ([1975] 1991) *The Tao of Physics* (3rd edn). Amerherst: Flamingo.

Capra, Fritjof (1988) *Uncommon Wisdom*. New York: Simon and Schuster.

Capra, Fritjof (1997) *The Web of Life: A New Synthesis of Mind and Matter.* London: Flamingo.

Carrington, Damian (2020) 'Halt the Destruction of Nature or Suffer Even Worse Pandemics, Warn World Scientists', *The Guardian*, 27 April.

Cesaire, Aime (1972) *Discourse on Colonialism*, trans. Joan Pinkham. London: Monthly Review Press.

Chacko, Priya (2016) 'The Decolonial Option: Toward an Ethic of Self-Securing', in Jonna Nyman and Anthony Burke (eds) *Ethical Security Studies: A New Research Agenda.*Oxfordshire: Routledge, pp 189–200.

Chai, David (2018) 'Rethinking the Daoist Concept of Nature', *Journal of Chinese Philosophy*, 43(3–4): 259–74.

Chan, Stephen, P. Mandaville and R. Bleiker (eds) (2001)*The Zen of International Relations*. London: Palgrave Macmillan.

Chan Wing-tsit (1963) *The Way of Lao Tzu*. New York: The Library of Liberal Arts, Bobbs-Merrill.

Chappel, Timothy (2005) *The Inescapable Self: An Introduction to Western Philosophy.* London: Weidenfeld and Nicolson.

Chatterjee, Amita (2018) 'In Search of Genuine Agency', in Sibesh Chandra Bhattacharya, Vrinda Dalmiya and Gangeya Mukherji (eds) *Exploring Agency in the Mahabharata: Ethical and Political Dimensions of Dharma*. Routledge India, pp 45–61.

Chen Min (1994) 'Sun Tzu's Strategic Thinking and Contemporary Business', *Business Horizons*, March-April.

Chen Xia and Martin Schonfeld (2011) 'A Daoist Response to Climate Change', *Journal of Global Ethics*, 7(2): 195–203.

Cheng Chung-Ying (1986) 'On the Environmental Ethics of Dao and the Ch'i', *Environmental Ethics*, 8(4): 251–370.

Cheng Chung-Ying (2018) 'Preface on Saving Anthropocene', *Journal of Chinese Philosophy*, 43(3–4): 175–7.

Cleary, Thomas (trans. & ed) (1989) *Mastering the Art of War: Zhuge Liang's and Liu Ji's Commentaries on the classic by Sun Tzu*. Boulder: Shambhala Dragon Editions.

Clegg, Brian (2006) *The God Effect: Quantum Entanglement, Science's Strangest Phenomenon*. New York: St. Martin's Griffin.

Coker, Christopher (2003) 'What would Sun Tzu say about the War on Terrorism?', *The RUSI Journal*, 148(1): 16–20.

Collins, Chuck (2021) 'Updates: Billionaire Wealth, U.S. Job Losses and Pandemic Profiteers', *Inequality*, [online] 26 January, available online from: https://inequality.org/great-divide/updates-billionaire-pandemic/

Collins, Tim (2010) *Behind the Lost Symbol*. New York: Penguin.

Compson, Jane (2014) 'Meditation, Trauma and Suffering in Silence: Raising Questions about how Meditation is Taught and Practiced in Western Contexts in Light of a Contemporary Trauma Resiliency Model', *Contemporary Buddhism*, 15(2): 274–97.

Connolly, William E. (1996) 'Suffering, Justice and the Politics of Becoming', *Culture, Medicine and Psychiatry*, 20: 251–77.

Coole, Diane (2005) 'Rethinking Agency: A Phenomenological Approach to Embodiment and Agentic Capacities', *Political* Studies, 53: 124–42.

Coole, Diane and Samantha Frost (eds) (2010) *New Materialisms: Ontology, Agency, and Politics.* Durham: Duke University Press.

Corneli, Alessandro (1987) 'Sun Tzu and the Indirect Strategy', *Revista di Studi Politici Internazionali*, 54(3): 419–45.

Cudworth, Erika and Stephen Hobden (2011) *Post-Human International Relations.* London: Zed Books.

Cudworth, Erika and Stephen Hobden (2013) 'Complexity, Ecologism and Post-Human Politics', *Review of International Studies*, 39(3): 643–64.

Dalai Lama XIV (2000) *Ancient Wisdom, Modern World: Ethics for a New Millennium*. London: Abacus.

Dalai Lama XIV (2005) *The Universe in a Single Atom: How Science and Spirituality Can Serve Our World*. London: Abacus.

Dasgupta, Partha (2021) *Final Report - The Economics of Biodiversity: The Dasgupta Review*. London: HM Treasury, available online from: https://www.gov.uk/government/publications/final-report-the-economics-of-biodiversity-the-dasgupta-review

Datta-Ray, Deep K. (2015) *The Making of Indian Diplomacy: A Critique of Eurocentrism*. London: Hurst & Co.

Dauphinee, Elizabeth (2010) 'The Ethics of Autoethnography', *Review of International Studies*, 36(3): 799–818.

Davidson, Richard J. and Cortland J. Dahl (2018) 'Outstanding Challenges in Scientific Research on Mindfulness and Meditation', *Perspectives on Psychological Science*, 13(1): 62–5.

Davidson, Richard and Anne Harrington (2002) *Visions of Compassion: Western Scientists and Tibetan Buddhists Examine Human Nature*. Oxford: Oxford University Press.

Davidson, Richard and Anthony Lutz (2008) 'Buddha's Brain: Neuroplasticity and Meditation', *IEEE Signal Process Mag*, 25(1): 174–6.

Davidson, Richard J., Jon Kabat-Zinn, Jessica Schumacher, Melissa Rosenkranz, Daniel Muller, Saki F. Santorelli, Ferris Urbanowski, Anne Harrington, Katherine Bonus and John F. Sheridan (2003) 'Alterations in Brain and Immune Function Produced by Mindfulness Meditation', *Psychosomatic Medicine*, 65: 564–70.

De Goede, Marieke and Gavin Sullivan (2016) 'The Politics of Security Lists', *Environment and Planning D: Society and Space*, 34(1): 67–88.

Delegard, Kirsten and Kevin Ehrman-Solberg (2017) '"Playground of the People"? Mapping Racial Covenants in 20th century Minneapolis', *Open Rivers: Rethinking Water, Place and Community*, 6, available online from: https://editions.lib.umn.edu/openrivers/article/mapping-racial-covenants-in-twentieth-century-minneapolis/

Deleuze, Gilles and Felix Guattari ([1987] 2016) *A Thousand Plateaus*. London: Bloomsbury.

Deligiannis, Tom (2012) 'The Evolution of Environment-Conflict Research: Towards a Livelihood Framework', *Global Environmental Politics*, 12(1): 78–100, available online from: https://www.mitpressjournals.org/doi/pdfplus/10.1162/GLEP_a_00098

Dennett, Daniel (1992) *Consciousness Explained*. Cambridge: MIT Press.

DerDerian, James (1987) *On Diplomacy: A Genealogy of Western Estrangement*. Hoboken: Wiley-Blackwell.

DerDerian, James and Alexander Wendt (2020) 'Quantizing International Relations: The Case for Quantum Approaches to International Theory and Security Practice', *Security Dialogue*, 51(5): 399–413.

Derk, Hans (2012) *History of the Opium Problem: The Assault on the East, c.a. 1600–1950*. Leiden: Brill.

Deschimaru Taisen (1982) *The Zen Way to the Martial Arts*. New York: Compass.

Dhand, Arti (2018) 'Kharmayo and the Vexed Moral Agent', in Sibesh Chandra Bhattacharya, Vrinda Dalmiya and Gangeya Mukherji (eds) *Exploring Agency in the Mahabharata: Ethical and Political Dimensions of Dharma*. Routledge India, pp 81–106.

Doidge, Norman (2007) *The Brain that Changes Itself*. London: Penguin.

Dolan, Kerry A. (2021) 'The Forbes World's Billionaires List: The Richest in 2021,' *Forbes*, 6 April, available online from: https://www.forbes.com/sites/kerryadolan/2021/04/06/forbes-35th-annual-worlds-billionaires-list-facts-and-figures-2021/?sh=3f6688ba5e58

Doran, Peter (2018) 'Mindfulness is just Buddhism Sold to you by Neoliberals', *The Independent*, 25 February.

Dorsey, Peter A. (2003) 'To "Corroborate Our Own Claims": Public Positioning and the Slavery Metaphor in Revolutionary America', *American Quarterly*, 55(3): 353–86.

Dudden, Alexis (2006) *Japan's Colonization of Korea: Discourse and Power*. Honolulu: University of Hawaii Press.

DuVernay, Ava (2016) *13th*, motion picture. California: Netflix.

Duyvendak, Jan Willem (2011) *The Politics of Home: Belonging and Nostalgia in Western Europe and the United States.* London: Palgrave.

Eacott, Jonathan (2016) *Selling Empire: India in the Making of Britain and America, 1600–1830.* Chapel Hill: UNC Press Books.

Easwaren, Eknath ([1985] 2007a) *The Bhagavad Gita.* Tomales: Nilgiri Press.

Easwaren, Eknath ([1985] 2007b) *The Dhammapada.* Tomales: Nilgiri Press.

Edkins, Jenny (2002) 'Forget Trauma? Responses to September 11', *International Relations*, 16(2): 243–56.

Edkins, Jenny (2006) 'Remembering Relationality: Trauma Time and Politics', in Duncan Bell (ed) *Memory, Trauma and World Politics: Reflections on the Relationship between Past and Present.* Basingstoke: Palgrave, pp 99–115.

Einstein, A., B. Poldolsky and N. Rosen (1935) 'Can a Quantum Mechanical Description of Physical Reality be Considered Complete', *Physical Review*, 47, available online from: http://cds.cern.ch/record/405662/files/PhysRev.47.777.pdf

Elliot, Larry (2017) 'Worlds Eight Richest People have the same wealth as the poorest 50%', *The Guardian*, 16 January.

Elliot, Larry and Damian Carrington (2021) 'Economics Failure over Destruction of Nature presents "Extreme Risks"', *The Guardian,* 2 February.

Epstein, Charlotte (2017) *Against IR Norms: Postcolonial Perspectives.* Oxfordshire: Routledge.

Eun Yong-Soo (2018) *What Is at Stake in Building "Non-Western" International Relations Theory?.* Oxfordshire: Routledge.

Eun Yong-Soo (2021) 'Calling for "IR as becoming-rhizomatic"', *Global Studies Quarterly*, 1(2): 1–24.

Fain, Kimberley (2017) 'The Devastation of Black Wall Street', *JSTOR Daily*, [online] 5 July, available online from: https://daily.jstor.org/the-devastation-of-black-wall-street/

Farooqui, Amar (2016) 'The Global Career of Indian Opium and Local Destinies', *Almanack*, 14: 52–73.

Fierke, K.M. (1998) *Changing Games, Changing Strategies: Critical Investigations in Security.* Manchester: Manchester University Press.

Fierke, K.M. (2012) *Political Self-Sacrifice: Agency, Body and Emotion in International Relations.* Cambridge: Cambridge University Press.

Fierke, K.M. (2015) *Critical Approaches to International Security.* London: Polity.

Fierke, K.M. (2017) 'Consciousness at the Interface: Wendt, Eastern Wisdom and the Ethics of Intra-action', *Critical Review*, 29(2): 141–69.

Fierke, K.M. (2019) '*Contrary sunt Complementa*: Global Entanglement and the Constitution of Difference', *International Studies Review*, 21(1): 146–69.

Fierke, K.M. and Francisco Antonio-Alfonso (2018) 'Language, Entanglement and the New Silk Roads', *Asian Journal of Comparative Politics*, 3(3): 194–206.

Fierke, K.M. and Vivienne Jabri (2019) 'Global Conversations: Relationality, Embodiment and Power on the road to a Global IR', *Global Constitutionalism*, 8(3), 506–35.

Fierke, K.M. and Nicola Mackay (2020) 'To See is to Break and Entanglement: Quantum Measurement, Trauma and Security', *Security Dialogue*, 51(5): 450–66.

Folger, Tim (2002) 'Does the Universe Exist if We are not Looking', *Discover*, 1 June.

Foo Check Teck (1997) *Reminiscences of an Ancient Strategist: The Mind of Sun Tzu*. Traverse City: Horizon.

Fox Richardson, Heather (2020) *How the South Won the Civil War.* Oxford University Press.

Framaran, Christopher (2018) 'The Theory of Karma in the Mahabharata', in Sibesh Chandra Bhattacharya, Vrinda Dalmiya and Gangeya Mukherji (eds) *Exploring Agency in the Mahabharata: Ethical and Political Dimensions of Dharma*. New Delhi: Routledge, pp 62–79.

Frescura, F.A.M. and Basil J. Hiley (1984) 'Algebras, Quantum Theory and Pre-space', *Revista Brasileira de Fiscia*, S2: S49–82.

Fung Yu-lan (trans) (1970) *Chuang Tzu*. New York: Gordan Press.

Gandhi, Mohandus ([1913] 1964) *The Collected Works of Mahatma Gandhi* (vol 12). New Delhi: The Publications Division, Ministry of Information and Broadcasting, Government of India.

Gandhi, Mohandus (1950) *Hindu Dharma*. Ahmedabad: Navajivan Publishing House.

Gandhi, Mohandus (1951) *Nonviolent Resistance (Satyagraha)*. New York: Schocken Books.

Gandhi, Mohandus (1999) *The Collected Works of Mahatma Gandhi* (vol 21). New Delhi: New Delhi Publications Division, Minstry of Information and Broadcasting, Government of India.

Gandhi, Mohandus (2009) *The Bhagavad Gita According to Gandhi*. Redford: Wilder Publications.

Gani, Jasmine (2017) 'The Erasure of Race: Cosmopolitanism and the Illusion of Kantian Hospitality', *Millennium*, 45(3): 425–66.

Ganeri, Jonardon (2012) *The Concealed Art of the Soul: Theories of Self and Practices of Truth in Indian Ethics and Epistemology.* Oxford: Oxford University Press.

Garfield, Jay (1994) 'Dependent Arising and the Emptiness of Emptiness: Why did Nāgārjuna Start with Causation?', *Philosophy East and West*, 44(2): 219–50.

Garfield, Jay (2001) 'Nāgārjuna's Theory of Causality: Implications Sacred and Profane', *Philosophy East and West*, 51(4): 507–24.

Garfield, Jay L. (2010) 'Taking Conventional Truth Seriously: Authority Regarding Deceptive Reality', *Philosophy East and West*, 60(3): 341–54.

Garikipati, Supriya and Uma Kambhampati (2021) 'Leading the Fight against the Pandemic: Does Gender "Really" Matter', *Feminist Economics*, 27(1–2): 401–18.

Geertz, Clifford (1968) *Islam Observed*. New Haven: Yale University Press.

Ghandnoosh, Nazgol (2015) 'Black Lives Matter: Eliminating Racial Inequity in the Criminal Justice System', *The Sentencing Project*, 3 February, aavailable online from: https://www.sentencingproject.org/publications/black-lives-matter-eliminating-racial-inequity-in-the-criminal-justice-system/

Gilbert, P. (2009) *The Compassionate Mind*. London: Constable.

Goleman, Daniel (2003) *Destructive Emotions: How Can We Overcome Them? A Scientific Dialogue with the Dalai Lama Narrated by Daniel Goleman*. New York: Bantam Books.

Gomez, Luis O. (1975) 'Some Aspects of the Free Will Question in the Nikayas', *Philosophy East and West*, 25: 81–90.

Goodale, Melvyn and David Milner (2013) *Sight Unseen: An Exploration of Conscious and Unconscious Vision*. Oxford: Oxford University Press.

Gossett, Thomas F. ([1963] 1997) *Race: The History of an Idea in America*. New York: Oxford University Press.

Grim, Patrick (1982) *Philosophy of Science and the Occult*. New York: SUNY Press.

Grovogui, Siba N. (2001) 'Come to Africa: A Hermeneutics of Race in International Theory', *Alternatives*, 26(4): 425–88.

Gyasi, Yaa (2016) *Homegoing*. New York: Knopf Publishing Group.

Hall, David L. (2001) 'From Reference to Deference: Daoism and the Natural World', in Norman J. Girardot, James Miller and Lui Xiaogan (eds) *Daoism and Ecology: Ways within a Cosmic Landscape*. Cambridge: Harvard University Press.

Hamilton, Scott (2017) 'Securing Ourselves *from* Ourselves? The Paradox of "Entanglement" in the Anthropocene', *Crime, Law, and Social Change*, 68: 579–95.

Hansen, Chad (1992) *A Daoist Theory of Chinese Thought*. Oxford: Oxford University Press.

Hansen, Chad (1994) '*Fa* (Standards: Laws) and Meaning Changes in Chinese Philosophy', *Philosophy East & West*, 44(3): 435–88.

Hansen, Chelsea (2019) 'Slave Patrols: An Early Form of American Policing', *Law Enforcement Museum*, [online] July 10, available online from: https://nleomf.org/slave-patrols-an-early-form-of-american-policing/

Hansen, Lene (2006) *Security as Practice: Discourse Analysis and the Bosnian War*. Oxfordshire: Routledge.

Hao Changchi (2006) '*Wu-Wei* and the Decentering of the Subject in Lao-Zhuang: An Alternative Approach in the Philosophy of Religion', *International Philosophical Quarterly*, 46(4): 445–57.

Harris, Cheryl (1993) 'Whiteness as Property', *Harvard Law Review*, 106(8): 1701–91.

Harrison, David M. (2000–2002) 'Complementarity and the Copenhagen Interpretation of Quantum Mechanics', Upscale, Department of Physics, University of Toronto, available online from: https://faraday.physics.utoronto.ca/PVB/Harrison/Complementarity/CompCopen.htm

Haven, Emmanuel and Andrei Krennikov (2013) *Quantum Social Science.* New York: Cambridge University Press.

Harvey, Paul (2007) 'Freedom of Willing in the Light of Theravada Buddhist Teachings', *Journal of Buddhist Ethics*, 14: 35–98.

Harvey, Peter (2000) *An Introduction to Buddhist Ethics.* Cambridge: Cambridge University Press.

Haymen, David (2018) 'Slavery: Scotland's Hidden Shame', *BBC Two Scotland*, 6 November, Series 1, Episode 1.

Hervia, James (2003) 'Opium, Empire and Modern History', *China Review International*, 10(2): 307–26.

Hesse, Herman (1951) *Siddhartha.* New York: Bantam Books.

Higate, Paul and Gurchathen Sanghera (2009) 'Positionality and Power: The Politics of Peacekeeping Research', *International Peacekeeping*, 16(4): 467–82.

Ho Ping-ti (2002) *Three Studies on Suntzu and Laotzu.* Taipei: Institute of Modern History, Academia Sinica.

Hobbes, Thomas ([1651] 1958) *Leviathan.* New York: Penguin Books.

Hobson, John (2004) *The Eastern Origins of Western Civilization.* Cambridge: Cambridge University Press.

Hochschild, Arlie (2016) *Strangers in the Own Land: Anger and Mourning on the American Right.* New York: The New Press.

Horgan, John (2018) 'David Bohm, Quantum Mechanics and Enlightenment', *Scientific American*, 23 July.

House of Commons (2020) 'Unequal Impact? Corona Virus and BAME People', Women and Equalities Committee, Third Report of Session, 2019–2021, available online from: https://committees.parliament.uk/publications/3965/documents/39887/default/

Howell, Alison (2012) 'The Demise of PTSD: From Governing through Trauma to Governing Resilience', *Alternatives: Global, Local, Political*, 37(3): 214–26.

Howell, Alison (2014) 'The Global Politics of Medicine: Beyond Global Health, Against Securitization Theory', *Review of International Studies*, 40: 961–87.

Hsu Funie (2016) 'What is the Sound of One Invisible Hand Clapping? Neoliberalism, the Invisibility of Asian and Asian American Buddhists, and Secular Mindfulness in Education', in Ronald E. Purser, David Forbes and Adam Burke (eds) *Handbook of Mindfulness.* Basel: Springer International Publishing, pp 369–381.

Hume, David ([1739–40] 1964) *A Treatise of Human Nature* (vol 1). London: Everyman.
Hyland, Terry (2014) 'Mindfulness, Free Will and Buddhist Practice: Can Meditation Enhance Human Agency?', *Buddhist Studies Review*, 31(1): 125–40.
Hyman, Jacques E.C. (2020) 'Symposium: Protean Power: Exploring the Uncertain and Unexpected in World Politics', *International Theory*, 12(3): 408–99.
Illumine, Nat (2019) *Whiteness: Deconstructed*, *Afropean*, [online] 12 May, available online from: https://afropean.com/whiteness-deconstructed/
Ince, Onur Ulas (2018) 'Between Commerce and Empire: David Hume, Colonial Slavery, and Commercial Incivility', *History of Political Thought*, 39(1): 107–34.
International Labour Organization (ILO) (2017) *Global Estimates of Modern Slavery: Forced Labour and Forced Marriage*. Geneva: International Labour Organization, available online from: https://www.ilo.org/global/publications/books/WCMS_575479/lang--en/index.htm
Ip Po-Keung (1983) 'Taoism and the Foundations of Environmental Ethics', *Environmental Ethics*, 5(4): 335–43.
Iyer, Raghavan (1973) *The Moral and Political Thought of Mahatma Gandhi*. New York: Oxford University Press.
Jabri, Vivienne (2012) *The Postcolonial Subject: Claiming Politics/Governing Others in Late Modernity*. London and New York: Routledge.
Jackson, Patrick Thaddeus and Daniel Nexon (1999) 'Relations before states: substance, process and the study of World Politics', *European Journal of International Relations*, 5(3): 291–332.
James, Simon P. (2004) *Zen Buddhism and Environmental Ethics*. Oxfordshire: Routledge.
Jerryson, Michael K. and Mark Juergensmeyer (2010) *Buddhist Warfare*. Oxford: Oxford University Press.
Jullien, Francois (2004) *A Treatise on Efficacy: Between Western and Chinese Thinking*. Honolulu: University of Hawaii Press.
Jullien, Francois (1995) *The Propensity of Things: Toward a History of Efficacy in China*. New York: Zone.
Kabat-Zinn, Jon (2005) *Coming to Our Senses: Healing Ourselves and the World through Mindfulness*. London: Piatkus.
Katzenstein, Peter J. and Lucia A. Seybert (2018) *Protean Power: Exploring the Uncertain and Unexpected in World Politics*. Cambridge: Cambridge University Press.
Kavalski, Emilian (2017) *The Guanxi of Relational International Theory*. Oxfordshire: Routledge.
Keeney, Bradford (2007) *Shaking Medicine: The Healing Power of Ecstatic Movement*. Rochester: Destiny Books.

Keohane, Robert (1988) 'International Institutions: Two Approaches', *International Studies Quarterly*, 32: 379–96.

Khoo Kheng-Hor (ed) (1992) *Sun Tzu's Art of War*. Selangor Darul Ehsan: Pelanduk Publications.

Kilomba, Grada (2010) *Plantation Memories: Episodes of Everyday Racism*. Munster: UNRAST -Verlag

Kim Hun Joon (2016) 'Will IR with Chinese Characteristics be a Powerful Alternative', *The Chinese Journal of International Politics*, 9(1): 59–79.

King, Gary, Robert Keohane and Sidney Verba (1994) *Designing Social Inquiry*. Princeton: Princeton University Press.

King, Robert James (2015) 'The Evolution of Non-action (*Wuwei*) in Daoism as Seen in the *Taipingjing*', *NII-Electronic Library Service*: 51–61, available online from: jstage.just.go.jp

King, Winston L. (1994) *Zen and the Way of the Sword: Arming the Samurai Psyche*. Oxford: Oxford University Press.

Kleine, Christoph (2006) '"The Epitome of the Ascetic life": The Controversy over Self-Mortification and Ritual Suicide as Ascetic Practices in East Asian Buddhism', in Oliver Freiburger (ed) *Asceticism and its Critics: Historical Accounts and Comparative Perspectives*. Oxford: Oxford University Press, pp 153–78.

Knightly, Nickilas (2013) 'The Paradox of Wuwei? Yes (and No)', *Asian Philosophy*, 23(2): 115–36.

Koschut, Simon (ed) (2020) *The Power of Emotions in World Politics*. Oxfordshire: Routledge.

Kovan, M. (2014) 'Thresholds of Transcendence: Buddhist Self-Immolation and Mahayanist Absolute Altruism', *Journal of Buddhist Ethics*, 20: 773–812.

Kratochwil, Friedrich (2018) *Praxis: On Acting and Knowing*. Cambridge: Cambridge University Press.

Krishna, Sankara (2015) 'A post-Colonial Racial/Spatial Order: Gandhi, Ambedkar and the Construction of the International', in Alexander Anievas, Nivi Manchanda and Robbie Shilliam (eds) *Race and Racism in International Relations: Confronting the Global Colour Line*. Oxfordshire: Routledge, pp 139–56.

Kuhn, Thomas ([1962] 1970) *The Structure of Scientific Revolutions*. Chicago: University of Chicago Press.

Kumar, Pradeep (2017) 'The Beautiful Integration of Martial Arts and Yoga', *Martial Arts Blog*,11 September, available online from: https://www.bookmartialarts.com/news/martial-arts-and-yoga

Kurki, Milja (2020) *International Relations in a Relational Universe*. Oxford: Oxford University Press.

Lai Karyn L. (2003) 'Conceptual Foundations for Environmental Ethics: A Daoist Perspective', *Environmental Ethics*, 25: 247–66.

Lai Karyn L. (2007) '*Ziran* and *Wuwei* in the *Daodejing*: An Ethical Assessment', *Dao*, 6: 325–37.

Lakoff, George and Mark Johnson (1980) *Metaphors We Live By*. Chicago: University of Chicago Press.

Lamb-Books, Benjamin (2016) '*Quantum Mind and Social Science*', *Perspectives: Newsletter of the Theory Section*, [online] 30 June, available online from: http://www.asatheory.org/current-newsletter-online/book-review-quantum-mind-and-social-science

Lao Tzu (1963) *Tao de Ching*. New York: Penguin Books.

Latour, Bruno (1996) *Aramis, or, The Love of Technology*. Cambridge: Harvard University Press.

Latour, Bruno (2004) *The Politics of Nature*. Cambridge: Cambridge University Press.

Lea, Jennifer, Louisa Cadman and Chris Philo (2014) 'Changing the Habits of a Lifetime? Mindfulness Meditation and Habitual Geographies', *Cultural Geographies*, 22(1): 49–65.

Lehr, Peter (2019) *Militant Buddhism: The Rise of Religious Violence in Sri Lanka, Myanmar and Thailand*. Basingstoke: Palgrave Macmillan.

Lepore, Jill (2018) *These Truths: A History of the United States*. New York: W.W. Norton.

Levine, Peter and Ann Federick (1997) *Waking the Tiger: Healing Trauma*. Berkeley: North Atlantic Books.

Li Zehou (2000) *Zhong Guo Gu Dai Si Xiang Shi Lun (On the History of Chinese Ancient Thought)*. Taipei: San Min Book, Co.

Li-Hua, Richard (2014) *Competitiveness of Chinese Firms: East Meets West*. London: Palgrave.

Lindsey, Peter, James Allan, Peadar Brehony, Amy Dickman, Ashley Robson, Colleen Begg, Hasita Bhammar, Lisa Blanken, Thomas Breuer, Kathleen Fitzgerald, Michael Flyman, Patience Gandiwa, Nicia Giva, Dickson Kaelo, Simon Nampindo, Nyambe Nyambe, Kurt Steiner, Andrew Parker, Dilys Roe, Paul Thomson, Morgan Trimble, Alexandre Caron and Peter Tyrrell (2020) 'Conserving Africa's wildlife and wildlands through the COVID-19 crisis and beyond', *Nature, Ecology and Evolution*, 29 July, available online from: https://www.nature.com/articles/s41559-020-1275-6

Ling, Lily H.M. (2013) 'Worlds beyond Westphalia: Daoist Dialectics and the China Threat', *Review of International Studies*, 39(3): 549–68.

Ling, Lily H.M. (2014) *The Dao of World Politics. Towards a Post-Westphalian Worldist International Relations*. London and New York: Routledge.

Little, Daniel (2018) 'Entangling the Social: Comments on Alexander Wendt, Quantum Mind and Social Science', *Journal of the Theory of Social Behavior*, 48: 167–76.

Liu Xiaogan (2001) 'Non-Action and the Environment Today: A Conceptual and Applied Study of Laozi's Philosophy', in Norman J. Girardot, James Miller and Lui Xiaogan (eds) *Daoism and Ecology: Ways within a Cosmic Landscape.* Harvard: Harvard University Press, pp 315–39.

Locke, John ([1690] 1978) *Two Treatises of Government.* New York: Everyman's Library.

Lopez, Donald S. (1995) *Buddhism in Practice.* Princeton: Princeton University Press.

Loy, David (1985) 'Wei-Wu-Wei: Nondual Action', *Philosophy East and West*, 35(1): 73–86.

Luo Tian and Zhang Meifang (2018) 'Reconstructing Cultural Identity via Paratexts: A Case Study on Lionel Giles' translation of The Art of War', *Perspectives*, 26(4): 593–611.

Macphail, Theresa (2014) *The Viral Network: A Pathography of the H1n1 Influenza Pandemic. Expertise: Cultures and Technologies of Knowledge.* Ithaca: Cornell University Press.

Malkso, Maria (2009) 'The Memory Politics of Becoming European: The East European Subalterns and the Collective Memory of Europe', *European Journal of International Relations*, 15(4): 653–80.

Mannion, A.M. (2003) *The Environmental Impact of War and Terrorism*, Geographical Paper No. 169, Reading: University of Reading.

Manual, Frank (1968) *A Portrait of Isaac Newton.* Cambridge: Belknap Press.

Marks, Robert (2015) *The Origins of the Modern World: A Global and Ecological Narrative.* Lanham: Rowman and Littlefield.

McLaughlin, Elliot C. (2020) 'America's Legacy of Lynching isn't all History. Many say it's still Happening Today', *CNN*, 3 June.

McMahan, David L. And Erik Braun (2017) 'From Colonialism to brain scans: Modern Transformations of Buddhist Meditation', in David McMahan and Erik Braun (eds) *Meditation, Buddhism and Science.* Oxford: Oxford University Press, pp 1–19.

McNally, Mark R. (2012) *Sun Tzu and the Art of Business: Six Strategic Principles for Managers.* Oxford: Oxford University Press.

Meister, Robert (1999) 'Forgiving and Forgetting: Lincoln and the Politics of National Recovery', in Carla Alison Hesse and Robert Post (eds) *Human Rights in Political Transitions: Gettysburg to Bosnia.* New York: Zone Books, pp 135–76.

Menakem, Resmaa (2017) *My Grandmother's Hands: Racialized Trauma and the Pathway to Mending our Hearts and Bodies.* Las Vegas: Central Recovery Press.

Meredith, Lisa S., Cathy D. Sherbourne, Sarah J. Gaillot, Lydia Hansell, Hans V. Ritschard, Andrew M. Parker and Glenda Wrenn (2011) *Promoting Psychological Resilience in the U.S. Military*. Santa Monica: Rand Corporation, available online from: https://www.rand.org/pubs/monographs/MG996.html

Meyer, Andrew Seth (2012) *The Dao of the Military: Liu An's Art of War.* New York: Columbia University Press.

Miles, Donna (2008) 'Centre Creates "Little Miracles" in Treating Combat Stress', *Defense Visual Information Distribution Service*, [online] 9 May, available online from: https://www.dvidshub.net/news/19278/center-creates-little-miracles-treating-combat-stress

Miller, James (2006) 'Daoism and Nature', in Roger S. Gottlieb (ed) *The Oxford Handbook of Religion and Ecology.* Oxford: Oxford University Press, pp 220–35.

Mills, China (2014) *Decolonising Global Mental Health: The Psychiatrization of the Majority World.* Oxfordshire: Routledge.

Mishra, Pankaj (2002) 'NS Essay – How the British Invented Hinduism', *New Statesman*, [online] 26 August, available online from: https://www.newstatesman.com/node/156145#:~:text=Yet%20Hinduism%20was%20a%2019th,Indus%20(Sindhu%20in%20Sanskrit)

Mitchell, Audra (2016) 'Posthuman Security/Ethics', in Jonna Nyman and Anthony Burke (eds) *Ethical Security Studies: A New Research Agenda.* London: Routledge, pp 60–72.

Mitra, Joy L. and Mark T. Greenberg (2016) 'The Curriculum of Right Mindfulness: The Relational Self and the Capacity for Compassion', in Ronald E. Purser, David Forbes and Adam Burke (eds) *Handbook of Mindfulness.* Geneva: Springer International Publishing, pp 411–24.

Moulds, Josephine (2021) 'Child Labour in the Fashion Supply Chain: Where, Why and What can be Done', *The Guardian*, [online] 19 January, available online from: https://www.theguardian.com/sustainable-business/ng-interactive/2015/jan/19/child-labour-in-the-fashion-supply-chain

Murphy, Michael P.A. (2020) *Quantum Social Theory for Critical International Relations Theorists: Quantizing Critique.* London: Palgrave Macmillan.

Nāgārjuna (1995) *The Fundamental Wisdom of the Middle Way: Nāgārjuna's Mulamadhyamakakarika.* Oxford: Oxford University Press.

Nandy, Ashis (2009[1983]) *The Intimate Enemy: Loss and the Recovery of Self Under Colonialism.* Oxford: Oxford University Press.

Ndlovu-Gatsheni, Sabelo J. (2018) *Epistemic Freedom in Africa: Deprovincialization and Decolonization.* Oxfordshire: Routledge.

Neihardt, John C. (2014) *Black Elk Speaks.* Lincoln: Bison Books.

Nelson, Eric Sean (2009) 'Responding with *Dao*: Early Daoist Ethics and the Environment', *Philosophy East and West*, 59(3): 294–316.

Ng, Edwin (2016) 'The Critique of Mindfulness and the Mindfulness of Critique: Paying Attention to the Politics of Our Selves with Foucault's Analytic of Governmentality', in Ronald E. Purser, David Forbes and Adam Burke (eds) *Handbook of Mindfulness.* Geneva: Springer International Publishing, pp 135–50.

Ngugi wa Thiong'o (2009) *Re-Membering Africa:* Nairobi, Kampala, Dar es Salaam: East African Educational Publishers.
Ngugi wa Thiong'o (2012) *Globalectics: Theory and Politics of Knowing.* New York: Columbia University Press.
Nhat Hanh (1999) *The Miracle of Mindfulness.* Boston: Beacon Press.
Nordin, Astrid, Graham M. Smith, Raoul Bunskoek, Huang Chiung-chiu, Hwang Yih-jye, Patrick Thaddeus Jackson, Emilian Kavalski, L. H. M. Ling (posthumously), Leigh Martindale, Nakamura Mari, Daniel Nexon, Laura Premack, Qin Yaqing, Shih Chih-yu, David Tyfield, Emma Williams and Marysia Zalewski (2019) 'Towards global relational theorizing: A Dialogue between Sinophone and Anglophone Scholarship on Relationalism', *Cambridge Review of International Affairs*, 32(5): 570–81.
Nottale, Laurent (1998) *Le Relativite Dans Tous Ses Etats.* Paris: Hachette.
Nyantiloka (1972) *A Buddhist Dictionary: Manual of Buddhist Terms and Doctrines.* Colombo: Frewin & Co., Ltd.
Nyman, Jonna and Anthony Burke (2016) *Ethical Security Studies: A New Research Agenda.* London: Routledge.
O'Brien, Karen (2016) 'Climate Change and Social Transformations: Is it Time for a Quantum Leap?', *WIREs Climate Change*, 7: 618–26.
O'Dowd, Edward and Arthur Waldron (1991) 'Sun Tzu for Strategists', *Comparative Strategy*, 10(1): 25–36.
Omnes, Roland (1999) *Quantum Philosophy: Understanding and Interpreting Contemporary Science.* Princeton: Princeton University Press.
Ooe Hiroaki (2014) 'Resonance frequency-retuned quartz tuning fork as a force sensor for noncontact atomic force microscopy', *Applied Physics Letters*, 105(4): 1–4.
Oppenheimer, Robert (1954) *Science and the Common Understanding.* Oxford: Oxford University Press.
Ostergaard, Geoffrey (1974) 'Gandhian Nonviolence and Passive Resistance', *Civil Resistance*, [online] 25 November, available online from: https://civilresistance.info/ostergaard
Oxfam (2021) 'Five Shocking Facts about Global Inequality and How to Even it Up', *Oxfam*, available online from: https://www.oxfam.org/en/5-shocking-facts-about-extreme-global-inequality-and-how-even-it
Pagels, Heinz R. (1982) *The Cosmic Code: Quantum Physics as the Language of Nature.* Mineola: Dover Books.
Pan Chengxin (2020) 'Enfolding Parts in Wholes: Quantum Holography and International Relations', *European Journal of International Relations*, 26(1): 14–38.
Pang-White, Anna A. (2016) 'Daoist *Ci*, Feminist Ethics of Care and the Dilemma of Nature', *Journal of Chinese Philosophy*, 43(3): 275–94.
Paquette, Laure (1991) 'Strategy and Time in Clausewitz's *On War* and Sun Tzu's *The Art of War', Comparative Strategy*, 10(1): 37–51.

Pasha, Mustapha Kamal (2017) *Islam and International Relations: Fractured Worlds*. Oxfordshire: Routledge.

Payutto, Bhikkhu P.A. (1993) *Good, Evil and Beyond: Kamma in the Buddha's Teaching*. Bangkok: Budhadhamma Foundation Publications.

Peat, F. David (2002) *Blackfoot Physics: A Journey in the Native American Universe*. Boston: Weiser.

Peerenboom, Randall P. (1991) 'Beyond Naturalism: A Reconstruction of Daoist Environmental Ethics', *Environmental Ethics*, 13: 3–22.

Perret, Ray (2002) 'Personal Identity, Minimalism and Madhyamaka', *Philosophy East and West*, 52(3): 373–85.

Petersen, Aage (1963) 'The Philosophy of Niels Bohr', *Bulletin of Atomic Scientists*, 19(7): 8–14.

Pigliucci, Massimo (2010) *Nonsense on Stilts: How to Tell Science from Bunk*. Chicago: University of Chicago Press.

Pilkingston, Ed (2020) 'UN Experts Condemn Modern-Day "Racial Terror" Lynchings in the U.S.', *The Guardian*, 5 June.

Planck, Max (1944) 'Das Wesen der Materie' ('The Nature of Matter'), speech in Florence, Italy, from Archiv zur Geschichte der Max-Planck-Gesellschaft, Abt. Va, Rep. 11 Planck, Nr. 1797.

Potter, Gary (2013) 'The History of Policing in the United States', *Police Studies Online*, Eastern Kentucky University, available online from: https://plsonline.eku.edu/insidelook/history-policing-united-states-part-1

Priest, Graham (2009) 'The Structure of Emptiness', *Philosophy East and West*, 59(4): 467–80.

Purser, Robert (2019) *McMindfulness: How Mindfulness Became the New Capitalist Spirituality*. London: Repeater Books.

Qin Yaqing (2007) 'Why is there no Chinese International Theory', *International Relations of the Asia-Pacific*, 7(3): 313–40.

Qin Yaqing (2011) 'Development of International Relations Theory in China: Progress Through Debates', *International Relations of the Asia-Pacific*, 11(2): 231–57.

Qin Yaqing (2012) 'Culture and Global Thought: Chinese International Theory in the Making', *Revisa CIDOB d'Afers Internacionals*, 100: 67–90.

Qin Yaqing (2016) 'A Relational Theory of World Politics', *International Studies Review*, 18(1): 33–47.

Qin Yaqing (2018) *A Relational Theory of World Politics*. Cambridge: Cambridge University Press.

Raju, Poola T. (1954) 'The Concept of the Spiritual in Indian Thought', *Philosophy East and West*, 4(3): 195–213.

Ravina, Mark (2004) *The Last Samurai: The Life and Battles of Saigo Takamori*. Hoboken: Wiley and Sons.

Rawls, John (1971) *A Theory of Justice*. Cambridge: Belknap Press.

Repetti, Rick (2012) 'Buddhist Hard Determinism: No Self, No Free Will, No Responsibility', *Journal of Buddhist Ethics*, 19: 130–97.

Repetti, Rick (2016) 'Meditation Matters: Replies to the Anti-McMindfulness Bandwagon!', in. Ronald E. Purser, David Forbes and Adam Burke (eds) *Handbook of Mindfulness.* Geneva: Springer International Publishing, pp 473–93.

Repetti, Rick (ed) (2019) *Buddhist Perspectives on Free Will: Agentless Agency?* London: Routledge.

Restak, Richard (1994) *The Module Brain: How Discoveries in Neuroscience are Answering Age Old Questions about Memory, Free Will, Consciousness and Personal Identity.* New York: Scribner.

Ricard, Matthew and Trinh XuanThuan (2001) *The Quantum and the Lotus: Where Science and Buddhism Meet.* New York: Three Rivers Press.

Ringmar, Eric (1996) 'On the Ontological Status of the State', *European Journal of International Relations*, 2(4): 439–66.

Roedinger, David (1991) *The Wages of Whiteness: Race and the Making of the American Working Class*. London: Verso.

Rosenbaum, Robert Meikyo and Barry Magid (eds) (2016) *What Is Wrong with Mindfulness (and What Isn't?): Zen Perspectives.* Somerville: Wisdom Publications.

Rosenfeld, Leon (1961) 'Foundations of Quantum Theory and Complementarity ', *Nature*, 190: 384–8.

Rosenfeld, Leon (1963) 'Niels Bohr's Contribution to Epistemology', *Physics Today*, 16(10): 47–54.

Rossdale, Chris (ed) (2015) 'Occupying Subjectivity: Being and Becoming Radical in the 21st Century', *Globalizations*, 12(1): 1–5.

Rovelli, Carlo (2018) *The Order of Time.* New York: Penguin Books.

Roy, Arundhati (2020) 'The Pandemic is a Portal', *Financial Times*, [online] 3 April, available online from: https://www.ft.com/content/10d8f5e8-74eb-11ea-95fe-fcd274e920ca

Said, Edward (1983) *World, the Text and the Critics.* Harvard: Harvard University Press.

Salter, Mark B. (2013) 'To Make Move and Let Stop: Mobility and the Assemblage of Circulation', *Mobilities*, 8(1): 7–19.

Salter, Mark B. (2015) *Introduction to Making Things International 1: Circuits and Motion*. Minneapolis: University of Minnesota Press.

Salter, Mark B. (2019) 'Security Actor-Network-Theory: Revitalizing Securitization Theory with Bruno Latour', *Polity*, 51(2): 349–64.

Sandel, Michael (2020) *The Tyranny of Merit: What's Become of the Common Good?* Bristol: Allen Lane.

Sanghera, Gurchathen S. and Suruchi Thapar-Björkert (2008) 'Methodological dilemmas: gatekeepers and positionality in Bradford', *Ethnic and Racial Studies*, 31(3): 543–62.

Sante Poromaa (2009) *The Net of Indra: Rebirth in Science and Buddhism.* Tallinn: Zendo.

Santos, Bouaventura de Sousa (2017) *Decolonising the University: The Challenge of Deep Cognitive Justice.* Cambridge: Cambridge Scholars Publishing.

Saunders, Robert (2020) 'Brexit and Empire: "Global Britain" and the Myth of Imperial Nostalgia', *The Journal of Imperial and Commonwealth History*, 48(6): 1140–74.

Sawyer, Ralph (1993) *The Seven Military Classics of Ancient China.* New York: Basic Books.

Sawyer, Ralph D. (trans) (1994) *Sun Tzu: The Art of War.* Boulder: Westview.

Sawyer, Ralph (1999) *The Dao of War.* New York: Basic Books.

Sawyer, Ralph (2007) *The Dao of Deception.* New York: Basic Books.

Schneider, Hans Julius (2017) 'Buddhist Meditation as Mystical Practice', *Philosophia*, 45: 773–87.

Schrodinger, Erwin (1944) *What is Life?* Cambridge: Cambridge University Press.

Scharf, Robert H. (2002) *Coming to Terms with Chinese Buddhism: A Reading of the Treasury Store Treatise.* Honolulu: University of Hawaii Press.

Settele, Josef, Sandra Diaz, Eduardo Brondizio and Peter Daszak (2020) 'Covid-19 Stimulus Measures Must Save Lives, Protect Livelihoods and Safeguard Nature to Reduce the Risk of Future Pandemics', *Inter Press Service*, 27 April, available online from: http://www.ipsnews.net/2020/04/covid-19-stimulus-measures-must-save-lives-protect-livelihoods-safeguard-nature-reduce-risk-future-pandemics/

Shanta, Bhakti Niskama (2015) 'Life and Consciousness – The Vedāntic View', *Communicative & Integrative Biology*, 8(5): 1–11.

Shay, Jonathan (2003) *Odysseus in America: Combat Trauma and the Trials of Homecoming.* New South Wales: James Bennett Pty Ltd.

Sheldrake, Rupert (2011) *Presence of the Past: Morphic Resonance and the Habits of Nature.* London: Icon Books.

Sheng-Ten, Ven Master (2000) *Setting in Motion the Dharma Wheel: Talks on the Four Noble Truths.* New York: Dharma Drum Publications.

Shermer, Michael (2005) 'Quantum Quackery', *Scientific American*, January.

Shih Chih-yu and Yin Jiwu (2013) 'Between Core National Interest and a Harmonious World: Reconciling Self-Role Conceptions in Chinese Foreign Policy', *The Chinese Journal of International Politics*, 6: 59–84.

Shih Chih-yu, Huang Chiung-chiu, Pichamon Yeophantong, Raoul Bunskoek, Josuke Ikeda, Hwang Yih-Jye, Wang Hung-Jen, Chang Chih-yun and Chen Ching-chan (2019) *China and International Theory: The Balance of Relationships.* London: Routledge.

Shilliam, Robbie (2011) *International Relations and Non-Western Thought.* London: Routledge.

Shilliam, Robbie (2020) 'Race and Racism in International Relations: Retrieving a Scholarly Inheritance', *International Politics Reviews*, 8: 152–95.
Shirer, William (1979) *Gandhi: A Memoir.* New York: Simon and Schuster.
Shotwell, Gordon (2013) 'Feeling Relational: The Use of Buddhist Meditation in Restorative Practices', *Dalhousie Law Journal*, 36(2): 317–34.
Siderits, Mark (2005) 'Freedom, Caring and Buddhist Philosophy', *Contemporary Buddhism*, 6: 87–116.
Siegel, Daniel J. (2007) *The Mindful Brain.* New York: W.W. Norton & Co.
Silove, Nina (2018) 'Beyond the Buzz Word: Three Meanings of "Grand Strategy"', *Security Studies*, 27(1): 27–57.
Singer, David (1961) 'The Level-of-Analysis Problem in International Relations', *World Politics*, 14(1): 77–92.
Singer, Joseph W. (1991) 'The Continuing Conquest: American Indian Nations, Property Law, and Gunsmoke', *Reconstruction*, 1(3): 97–104.
Singer, Peter (2004) *One World: The Ethics of Globalization* (2nd edn). New Haven: Yale University Press.
Sjoberg, Laura (2020) 'Quantum Ambivalence', *Millennium*, 49(1): 126–39.
Slingerland, Edward (2000) 'Effortless Action: The Chinese Spiritual Ideal of Wu-wei', *Journal of the American Academy of Religion*, 68(2): 293–327.
Slomp, Gabriella (2018) 'Hobbes on Benevolence and Love of Others', in Sharon Lloyd (ed) *Interpreting Hobbes' Political Philosophy.* Cambridge: Cambridge University Press, pp 106–23.
Smetham, Graham (2010) *Quantum Buddhism: Dancing in Emptiness.* Washington: Shunyata Press.
Smithsonian National Museum of African American History and Culture (2019) 'This Deplorable Entanglement', *Monticello*, available online from: https://www.monticello.org/slavery/paradox-of-liberty/thomas-jefferson-liberty-slavery/this-deplorable-entanglement/
Snyder, Samuel (2006) 'Chinese Traditions and Ecology: Survey Article', *Worldviews*, 10(1): 100–34.
Soeng Mu (2004) *Trust in Mind: The Rebellion of Chinese Zen.* Boston: Wisdom Publications.
Song Xinning (2001) 'Building International Theory with Chinese Characteristics', *Journal of Contemporary China*, 10(26): 61–74.
Stapp, Henry (1971) '*S*-Matrix Interpretation of Quantum Theory', *Physical Review D*, 3(6): 1303–20.
Stapp, Henry (2007) *Mindful Universe.* Berlin, Heidelberg: Springer-Verlag.
Statista (2020) 'Number of novel coronavirus (COVID-19) deaths of selected countries in the Asia Pacific region as of December 13, 2020', *Statista*, available online from: https://www.statista.com/statistics/1104268/apac-covid-19-deaths-by-country/
Steele, Brent (2008) *Ontological Security in International Relations: Self Identity and the IR State.* London: Routledge.

Steele, Brent (2019) *Restraint in International Politics.* Cambridge: Cambridge University Press.

Steger, Manfred B. (2006) 'Searching for *Satya* through *Ahimsa:* Gandhi's Challenge to Western Discourses of Power', *Constellations,* 13(3): 332–53.

Stenger, Victor J. (1997) '"Quantum Quackery"', *Skeptical Inquirer,* 21(1): 37–42.

Stenholm, Stig (2011) *The Quest for Reality: Bohr and Wittgenstein, Two Complementary Views.* Oxford: Oxford University Press.

Sun Tzu and Rim Sun (2007) *The Complete Art of War.* New York: Basic Books.

Suzuki Diasetsu Teitaro (1963) *Outlines of Mahayana Buddhism.* New York: Schocken Books.

Sylvester, Christine (2005) 'The Art of War/The War Question in (Feminist) IR', *Millennium,* 33(3): 855–78.

Szabo, Sandor P. (2003) 'The Term Shenming – Its Meaning in the Ancient Chinese Thought and in a Recently Discovered Manuscript', *Acta Orientalia Academiae Scientiarum Hungaricae,* 56: 251–74

Taleb, Nassim Nicholas (2010) *The Black Swan: The Impact of the Highly Improbable.* New York: Penguin Books.

Tang Yi-Yuan and Michael I. Posner (2012) 'Introduction', *Social Cognitive and Affective Neuroscience,* 8(S1), S1–3.

Tao Jiang (2014) 'The Incommensurability of Two Conceptions of Reality: Dependent Origination and Emptiness in Nāgārjuna's *MMK*', *Philosophy East and West,* 64(1): 25–48.

Tegmark, Mark (2015) 'Consciousness is the way information feels', *Church of the Churchless,* [online] 3 December, available online from: https://hinessight.blogs.com/church_of_the_churchless/2015/12/max-tegmark-consciousness-is-the-way-information-feels.html

Temperton, James (2017) 'I am become Death the Destroyer or Worlds: The Story of Oppenheimer's Famous Quote', *Wired,* [online] 8 September, available online from: https://www.wired.co.uk/article/manhattan-project-robert-oppenheimer

Thompson, Evan (2017) 'Looping Effects and the Cognitive Science of Mindfulness Meditation', in David McMahan and Erik Braun (eds) *Meditation, Buddhism and Science.* Oxford: Oxford University Press.

Tilley, Lisa and Robbie Shilliam (eds) (2018) 'Raced Markets', *New Political Economy,* 23(5): 531–639.

Titmuss, Christopher (2016) 'Is there a Corporate Takeover of the Mindfulness Industry? An Exploration of Western Mindfulness in the Public and Private Sector', in Ronald E. Purser, David Forbes and Adam Burke (eds) *Handbook of Mindfulness.* Geneva: Springer International Publishing, pp 181–94.

TMI (The Mindfulness Initiative) (2019) 'The Story So Far', *The Mindfulness Initiative*, available online from: https://www.themindfulnessinitiative.org/story-so-far

Tognini, Geocomo (2021) 'The Countries with the Most Billionaires in 2021', *Forbes*, 6 April, available online from: https://www.forbes.com/sites/giacomotognini/2021/04/06/the-countries-with-the-most-billionaires-2021/?sh=7785e7c379b2

Treckner, V. (ed) (1935) *Majjhima Nikaya*. London: Pali Text Society.

Tuhiwai-Smith, Linda (2012) *Decolonizing Methodologies: Research and Indigenous Peoples.* London: Zed Books.

UNHCR (2020) 'Forced Displacement passes 80 million by mid-2020 as COVID-19 tests refugee protection globally', *UNHCR*, [online] 9 December, available online from: https://www.unhcr.org/uk/news/press/2020/12/5fcf94a04/forced-displacement-passes-80-million-mid-2020-covid-19-tests-refugee-protection.html

Uzgalis, William (2018) 'Notes to John Locke', *Stanford Encyclopedia of Philosophy*, available online from: https://plato.stanford.edu/entries/locke/notes.html

Vago, David R. and David A. Silbersweig (2012) 'Self-Awareness, self-regulation, and self-transcendence (S-ART): A framework for understanding the neurobiological mechanisms of mindfulness', *Frontiers in Human Neuroscience*, 25: 1–58.

Valera, Francisco J., Evan Thompson and Eleanor Rosch ([1991] 2017) *The Embodied Mind: Cognitive Science and Human Experience.* Cambridge: MIT Press.

Van Daele, Wim (2018) 'Food as the Holographic Condensation of Life in Sri Lankan Rituals', *Ethnos*, 83(4): 645–54.

Van Norden, Bryan (2008) *Mengzi, with Selections from Traditional Commentaries*. Indianapolis: Hackett Publishing.

Verlet, Loup (1993) *La Malle de Newton.* Paris: Gallimard.

Vetter, Tilmann (1988) *The Ideas and Meditative Practices of Early Buddhism.* Leiden: Brill.

Vincent, Raymond J. (1988) *Human Rights and International Relations.* Cambridge: Cambridge University Press.

Vitale, Christopher (2014) *Networkologies: A Philosophy of Networks for a Hyperconnected Age.* Ropley: John Hunt Publishing.

Voelkner, Nadine (2019) 'Riding the Shi: From Infection Barriers to the Microbial City', *International Political Sociology*, 13(4): 376–91.

Waldman, Felix (2014) *Further Letters of David Hume.* Edinburgh: Edinburgh Bibliographical Society.

Waldner, David (2017) 'Schrodinger's Cat and the Dog that Didn't Bark: Why Quantum Mechanics is (Probably) Irrelevant to the Social Sciences', *Critical Review*, 29(2): 199–233.

Waldron, William S. (2002) 'Buddhist Steps to an Ecology of Mind: Thinking about "Thoughts without a 'Thinker'"', *The Eastern Buddhist*, 34(1): 1–52.

Waldron, William S. (2017) 'Reflections on Indian Buddhist Thought and the Scientific Study of Meditation, or Why Scientists should talk more with the Buddhist Subjects', in David McMahan and Erik Braun (eds) *Meditation, Buddhism, and Science*. Oxford: Oxford University Press, pp 1–29.

Waley, Arthur (1958) *The Way and Its Power: A Study of the Tao Te Ching and Its Place in Chinese Thought*. New York: Grove Press.

Walker, Sam (2020) 'Thank God for Calm, Competent Deputies', *The Wall Street Journal*, [online] 4 April, available online from: https://www.wsj.com/articles/in-the-coronavirus-crisis-deputies-are-the-leaders-we-turn-to-11585972802

Walsh, Zack (2016) 'A Meta-Critique of Mindfulness Critiques: From McMindfulness to Critical Mindfulness', in Ronald E. Purser, David Forbes and Adam Burke (eds) *Handbook of Mindfulness*. Geneva: Springer International Publishing, pp 153–66.

Wang Chen (1999) *The Tao of Peace: Lessons from Ancient China on the Dynamics of Conflict*. London: Shambala.

Wang Hung-Jen (2013) *The Rise of China and Chinese International Relations Scholarship*. Washington: Lexington Books.

Wang Hung-Jen (2017) 'Traditional empire–modern state hybridity: Chinese Tianxia and Westphalian anarchy', *Global Constitutionalism*, 6(2): 298–326.

Wang, Robin R. (2012) *Yinyang: The Way of Heaven and Earth in Chinese Thought and Culture*. Cambridge: Cambridge University Press.

Wang Yiwei (2007) 'Between Science and Art: Questionable International Relations Theories', *Japanese Journal of Political Science*, 8(2): 191–208.

Wanjek, Christopher (2016) 'Primates, including Humans, are the Most Violent Animals', *LiveScience*, [online] 28 September, available online from: https://www.livescience.com/56306-primates-including-humans-are-the-most-violent-animals.html

Warner, Michael (2006) 'The Divine Skein: Sun Tzu on Intelligence', *Intelligence and National Security*, 21(4): 483–92.

Watters, Ethan (2010) *Crazy Like Us: The Globalization of the American Psyche*. London: Free Press.

Weber, Max ([1919] 2004) *'Politics as a Vocation', The Vocation Lectures*. Illinois: Hackett Books.

Weber, Max ([1958] 1976) *The Protestant Ethic and the Spirit of Capitalism*. Crows Nest: George Allen and Unwin Publishers.

Weekley, Ernest (1967) *An Etymological Dictionary of Modern English*. New York: Dover.

Wendt, Alexander (1992) 'Anarchy is What States Make of it', *International Organization*, 40(2): 391–425.

Wendt, Alexander (2004) 'The State as a Person in International Relations', *Review of International Studies*, 30(2): 289–316.

Wendt, Alexander (2015) *Quantum Mind and Social Science.* Cambridge: Cambridge University Press.
Westfall, Richard ([1980] 1993) *Newton.* Cambridge: Cambridge University Press.
Westing, Arthur H. (2012) *Pioneer on the Environmental Impact of War.* New York: Springer.
Wiener, Antje (2018) *Contestation and Constitution of Norms in Global International Relations.* Cambridge: Cambridge University Press.
Wienpahl, Paul (1979) 'Eastern Buddhism and Wittgenstein's Philosophical Investigations', *The Eastern Buddhist*, 12(2): 22–54.
Wight, Colin (2018) 'Commentary: Forum Introduction', *Journal for the Theory of Social Behaviour*, 48: 154–56.
Williams, Caroline (2002) 'The Subject and Subjectivity', in Alan Finlayson and Jeremy Valentine (eds) *Politics and Post-structuralism: An Introduction.* Edinburgh: Edinburgh University Press, pp 23–35.
Williams, Mark (2011) *Mindfulness: A Practical Guide to Finding Peace in a Frantic World.* London: Piatkus.
Wittgenstein, Ludwig (1922) *Tractatus Logico-Philosophicus.* London: Kegan Paul.
Wittgenstein, Ludwig (1958) *Philosophical Investigations.* Hoboken: Blackwell
Yamamoto Tsunetomo (2000) *Hagakure: The Book of the Samurai.* London: Kodansha Europe.
Yan Xuetong (2011) *Ancient Chinese Thought, Modern Chinese Power.* Princeton: Princeton University Press.
Yuen, Derek M.C. (2008) 'Deciphering Sun Tzu', *Comparative Strategy*, 27(2): 183–200.
Yuen, Derek M.C. (2014) *Deciphering Sun Tzu: How to Read the Art of War.* London: Hurst & Co.
Zalewski, Marysia (2019) 'Forget(ting) Feminism? Investigating Relationality in IR', *Cambridge Review of International Studies*, 32(5): 615–35.
Zanotti, Laura (2019) *Ontological Entanglements, Agency and Ethics in International Relations: Exploring the Crossroads.* London: Routledge.
Zhang Feng (2012) 'The Tsinghua Approach and the Inception of Chinese Theories of International Relations', *The Chinese Journal of International Politics*, 5(1), 73–102.
Zhang Yongjin and Chang Ten-Chi (2016) *Constructing a Chinese School of International Relations.* London: Routledge
Zhao Tinyang (2006) 'Rethinking Empire from a Chinese Concept "All-under-heaven" (Tianxia)', *Social Identities*, 12(1): 29–41.
Zhao Tingyang (2009) 'A Political World Philosophy in terms of All-under-heaven (Tian-xia)', *Diogenes*, 56(1): 5–18.
Zhao Tinyang (2021) *All Under Heaven: The Tianxia System for a Possible World Order.* Los Angeles: University of California Press.

Ziporyn, Brook (2009) *Zhuangzi: The Essential Writings.* Cambridge: Hackett Publishing.

Zizek, Slavoj (2001) 'From Western Marxism to Western Buddism', *Cabinet Magazine*, 2, available online from: http://www.cabinetmagazine.org/issues/2/western.php

Zizek, Slavoj (2012) 'The Buddhist Ethic and the Spirit of Global Capitalism', lecture, European Graduate School, 25 August, transcript available online from: https://zizek.uk/the-buddhist-ethic-and-the-spirit-of-global-capitalism/

Zohar, Danah and Ian Marshall (1994) *The Quantum Society: Mind, Physics and a New Social Vision.* Amherst: Flamingo.

Zwick, Edward (2003) *The Last Samurai*, motion picture, Los Angeles: Radar Pictures and Warner Brothers.

Index

References to endnotes show both the page number and the note number (231n3).